CONTENTS

INTRODUCTION

The ancient Romans helped to shape the world as we know it today. The rise and ultimate fall of their civilization is central to the early history of Europe and the Mediterranean basin, and their cultural and political influence has been profound over succeeding centuries.

The Roman republic inherited and built upon the greatness of ancient Greece, absorbing Greek advances in fields ranging from philosophy and religion to science and the arts. The Roman empire, through its military occupation of much of western Europe, the Middle East and north Africa, ensured the communication of such ideas to the rest of the known world.

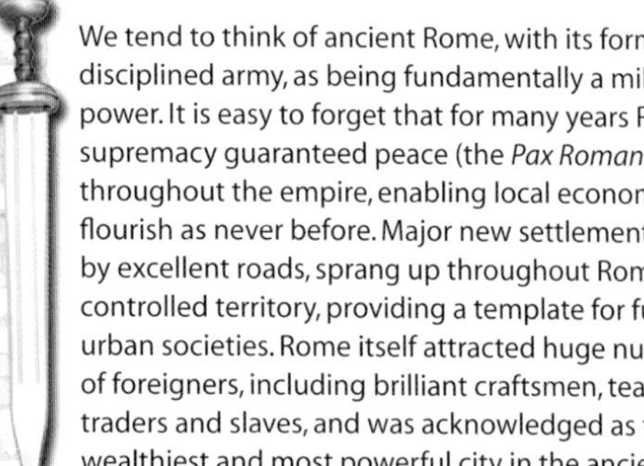

We tend to think of ancient Rome, with its formidable, disciplined army, as being fundamentally a military power. It is easy to forget that for many years Roman supremacy guaranteed peace (the *Pax Romana*) throughout the empire, enabling local economies to flourish as never before. Major new settlements, linked by excellent roads, sprang up throughout Roman-controlled territory, providing a template for future urban societies. Rome itself attracted huge numbers of foreigners, including brilliant craftsmen, teachers, traders and slaves, and was acknowledged as the wealthiest and most powerful city in the ancient world.

The influence of Rome spread as far as Spain in the west, Syria in the east, the Sahara in the south and Britain in the north. For 1000 years, Rome was the world's leading power and its intellectual and commercial centre.

ROME AND THE MODERN WORLD

The Romans left behind many spectacular ruins, such as temples, amphitheatres, aqueducts and roads. Even more significant was their legacy of ideas. Among the innovations bequeathed by the Romans to later civilizations were not only their history and mythology but also the Roman alphabet and system of numerals, the Roman calendar, the use of concrete, glass windows and domes, central heating, a public health system, public baths, hospitals, a postal service, a fire brigade, a civil service, apartment blocks, international trade and the idea of empire. The Romans also had an enduring impact upon politics (many of their frontiers becoming the borders of modern states), law, religion (not least through the formation of the Roman Catholic Church) and language.

A united Europe

Many historians trace the modern ideal of a united Europe, with a shared government, linked economies and system of international law, back to ancient Rome.

PART ONE

The land of the Romans

Roman territory extended far beyond Rome itself, coming to include, in due course, the entire Italian peninsula, the coastline of the Mediterranean and most of what is now western Europe, as well as territories in northern Africa and the Middle East. Conquered lands were made Roman provinces, thus ensuring that the Roman way of life had a long-lasting and widespread influence.

View of Ancient Rome

For hundreds of years, Rome, to which all roads were said to lead, was the most magnificent city in the ancient world.

THE ROMAN WORLD

Italy itself has a varied landscape, ranging from fertile plains and marshes to the rocky mountainsides of the Apennines running down the central spine of the country. Some of the most fertile areas are located on the west coast, and it was here that the Roman civilization first emerged on the banks of the Tiber, which provided easy access to the coast, some twenty-five kilometres (six miles) downstream. Later Roman prosperity was to depend greatly upon Italy's central position within the Mediterranean basin and trade routes with other peoples.

NEIGHBOURING PEOPLES

Early Roman expansion was at the expense of such neighbouring peoples as the Sabines and the Etruscans. This was a prolonged process, with Rome itself often falling victim to invasion by hostile tribes before Roman supremacy was eventually established throughout central Italy. Most of the kings of Rome, in fact, were of Sabine or Etruscan origin.

Rome's contact with such neighbouring peoples, through trade and many other transactions as well as through warfare, was to play a crucial part in the early development of Roman culture.

The Roman world *c.* 200 BC

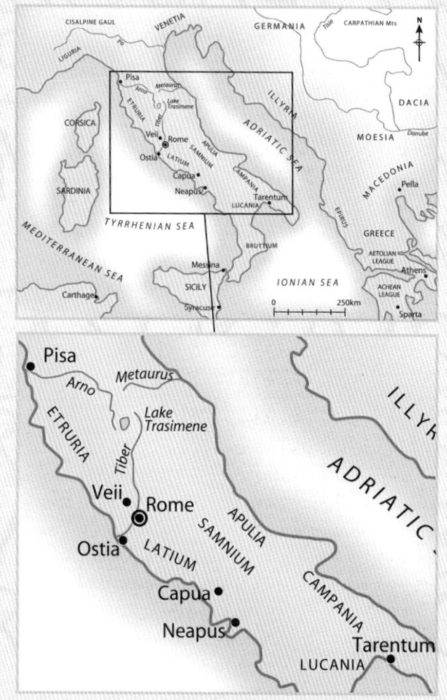

THE ROMAN EMPIRE

All of Italy was under Roman control by the middle of the third century (264 BC), after which Roman armies ventured further afield, seizing much of Hispania (modern Spain) from Rome's rival Carthage and bringing Macedonia under its influence by 168 BC. In 133 BC, Pergamum became the first of Rome's provinces in Asia. Gallia Narbonensis (southern Gaul) was made a Roman territory in 121 BC, while Cyrene in north Africa became the Roman province of Cyrenaica in 96 BC. The conquest of Bithynia, Pontus, Syria and Crete between 75 BC and 64 BC extended Rome's possessions in the eastern Mediterranean.

EXPANSION

Julius Caesar had conquered the rest of Gaul by 49 BC and even ventured north into Britain, while Egypt was added by Octavian in 31 BC. Rome's territories continued to grow after the foundation of the empire under Augustus, who put down rebellions and established Roman rule over all of mainland Europe west and south of the Rhine and Danube rivers. The Roman empire reached its further extent in the early years of the second century AD under the emperor Trajan, who successfully conquered Dacia (modern Romania) and large parts of the Middle East.

Frontiers

Roman influence was not confined to the borders of the empire. Roman troops often penetrated far beyond frontiers into lands held by hostile peoples in order to neutralize the threat of invasion, to support allies of Rome or to defend trading or other interests. The borders themselves were not necessarily strongly fortified. Hadrian's Wall, with its mighty wall and linked forts, was the most heavily defended frontier, and the German frontier was similarly protected by a timber wall and forts, but elsewhere there were relatively light border fortifications, if any at all.

Disintegration of the empire

By now too large to be governed effectively, the empire slowly contracted after Trajan's reign and was split into two – with rival capitals in Rome and Constantinople – and eventually disintegrated after repeated civil wars and barbarian invasions.

The empire revived

Memories of the empire were revived in the sixth century AD when Justinian, emperor of the eastern empire, reconquered much of Rome's former territory around the Mediterranean and in southern Europe. However, the cost of running this rebuilt empire proved impossible to meet, and these regions soon returned to rule by barbarian kings.

The Roman empire

Oceanus
Germanicus
Britannia
GERMANIA
Germania
Inf.
Belgica
Lugdunensis
Germania
Sup.
Rhaetia
Noric
Oceanus
Atlanticus
GALLIA
Aquitania
Gallia
Cisalpina
Narbonensis
Gataecia et
Asturia
ITALIA
Rome
Corsica
Tarraconesis
HISPANIA
Sardinia
Mare
Tyrrhanum
Lusitania
Balaeres
Baetica
Sici
Numidia
Mauretania
Africa
Proconsularis
0
500km

N
SARMATIA
nnonia
Dacia
LYRIUM
Dalmatia
Moesia Sup.
Moesia Inf.
Pontus Euminos
Thracia
Macedonia
Bithynia
Pontus
ARMENIA
Epirus
ASIA
Galatia
Cappadocia
Achaea
Phrygia
Mare Aegaeum
Lycia
Cilicia
Syria
MARE INTERNUM
MESOPOTAMIA
Palaestina
Cyrenaica
Aegyptus
Arabia

ROMAN PROVINCES

Rome organized conquered territories as provinces, each under the rule of a Roman governor of senatorial rank. The number and size of Rome's provinces varied over the centuries, some regions enjoying greater peace and prosperity than others.

AFRICA

In 146 BC, having defeated Carthage in the second Punic war, the Romans founded the province of Africa (their first on the African continent), which was roughly equivalent to modern northern Tunisia and western Libya and included the remains of Carthage itself. Later, Roman rule was extended to include the entire north African coastline, from the province of Mauretania (now western Algeria and northern Morocco) in the west to Aegyptus (modern Egypt) in the east. The importance of the region lay not only in its strategic influence, but

Roman towns

Many British towns have Roman origins. The most important include Bath (Aqua Sulis), Canterbury (Durovernum), Colchester (Camulodonum), Chester (Deva), St Albans (Verulamium) and York (Eboracum).

also in its massive agricultural output, which included cereals, fruit and olive oil. It was also a rich source of slaves and wild animals for Rome's circuses.

ASIA

Rome's first province in Asia was Pergamum, which was acquired in 133 BC. Later military campaigns added much more territory to the east and the south, with the addition of the province of Judaea linking with Roman territories in northern Africa and thereby completing Roman control of the entire Mediterranean coastline. Important cities in the region included Ephesus, Antioch and Byzantium which, as Constantinople, became the capital of the Roman empire in 326 AD.

BRITANNIA

The Roman legions first visited Britain in 55 BC and 54 BC under Julius Caesar, attracted by tales of tin mines and other riches, but they only established a more permanent presence with the second invasion of 43 AD, occupying the mainland as far north as the border of Caledonia until 410 AD. Having landed at Richborough in Kent, the legions won a series of victories against the Catuvellauni tribe and their allies and captured their leader Caratacus, who was sent to Rome to be presented to the emperor Claudius (who was impressed by his bearing and spared his life).

The Roman humiliation of Boudicca sparked a revolt that nearly drove the Romans from Britain.

Rebellions

Roman rule was disrupted by rebellions, most notably the Iceni revolt of 61 AD led by Queen Boudicca, which led to the sacking of Londinium (modern London) and several other towns before ending in defeat in battle against the legions and the suicide of Boudicca. The Caledonians were defeated by the governor Agricola at Mons Graupius in 84 AD, and the threat of invasions from northern barbarians was warded off from 120 AD by the building of Hadrian's Wall, which connected northern Britain's west and east coasts. A second wall, the Antonine Wall, was built further north circa 142 AD following reoccupation of the Scottish lowlands.

The Romans gradually withdrew from Britannia, their northernmost province, in the early fifth century. Their legacy included roads, villas, forts, bath complexes and towns as well as advances in agriculture and industry.

GAUL

Gaul (or Gallia) was the name given by the Romans to a large area of western Europe now divided between northern Italy, France, Belgium and western Switzerland as well as parts of the Netherlands and Germany. The Romans established their first province on the northern side of the Alps in 121 BC, and went on to subdue the Celtic and Germanic peoples of the entire region under Julius Caesar in the years 58–49 BC, with a final victory over the Gaulish army of Vercingetorix at Alesia (52 BC).

Gaulish provinces

The Gaulish territories were reorganized by the Romans into several provinces. Gallia Aquitania encompassed what is now southwest France, while southeast France became Gallia Narbonensis, central and northwest France became Gallia Lugdunensis, and the northeastern parts were split between Gallia Belgica, Germania Inferior and Germania Superior. These divisions are reflected today in the national borders of western Europe. Important Roman towns within Gaul included Lugdunum (modern Lyons), which was the capital of Gallia Lugdunensis, Lutetia (modern Paris), Massalia (modern Marseilles) and Narbo Martius (Narbonne).

GREECE

Rome established control of Greece and neighbouring territories in 146 BC at the end of a bloody campaign that culminated in the destruction of the city of Corinth and the massacre of its inhabitants. A later rebellion was put down by Lucius Cornelius Sulla, resulting in the sacking of Athens and Thebes. The Aegean islands were added in 133 BC. The region erupted into violence several times, notably during the revolt of the Greek cities in 88 BC and again during the Roman civil wars of the late republican era. Augustus reorganized Greece as the province of Achaea in 27 AD, while neighbouring states subsequently became the provinces of Thrace, Macedonia, Illyricum, Moesia and Pannonia.

The cultural links between Rome and the Greek world were long-standing and, even within the Roman empire, the Greeks continued to exert a profound influence upon Roman civilization. Many Greek slaves were brought to Rome to work as doctors and teachers and the area also provided soldiers for Roman armies.

Vercingetorix, the commander of the Gaulish army, surrendered to Julius Caesar in 52 BC.

HISPANIA

After being for many years under Carthaginian control, much of what is now Spain finally fell to Rome in 206 BC, following Rome's victories in the Punic Wars. The Romans split the area into two and, later, three provinces: Hispania Baetica (southern Spain), Hispania Lusitania (roughly equivalent to modern Portugal and part of western Spain) and Hispania Tarraconensis (the north and east of the country). Hispania was a great source of wealth, producing grain, silver, wine and olive oil.

THE CITY OF ROME

The city of Rome developed from a group of villages situated on seven hills on the plain of Latium beside the River Tiber, on the borders of Etruria. The marshy valleys surrounding the hills made the area unhealthy, but the hills provided a strong defensive position. The Tiber could be crossed by a ford at this point, which also marked the furthest inland point that ships could reach.

THE FORUM

As Rome grew, a marketplace known as the Forum Romanum was established on a patch of drained marshland and then paved. In due course this became the site of many of the city's most important buildings (including temples and the Curia, where the Senate met) and the focus of city life, where citizens gathered to meet their friends and hear the news. Through the Forum ran the Via Sacra (Sacred Way) leading to the Capitoline Hill and the Temple of Jupiter. Distances in Italy were traditionally measured from Rome's Forum.

Over the centuries, Rome continued to be enlarged and rebuilt many times, with the addition of temples, new forums, imperial palaces, apartment blocks (*insulae*), warehouses, shops, baths, amphitheatres, triumphal arches, aqueducts and other structures, many of them built using towering cranes. Many emperors sought to achieve lasting fame through their improvements to the city. Augustus, for instance, boasted that he 'found Rome built of bricks and left her covered in marble'.

REBUILDING

Rome was badly damaged by fire on several occasions, most notably in the Great Fire of 64 AD, after which it was rebuilt on an ambitious scale by the emperor Nero. Building projects continued under such emperors as Vespasian, who began the Colosseum, and Titus, builder of the Arch of Titus celebrating the defeat of the Jewish revolt of 73 AD. Although there were poor quarters, many buildings were richly decorated with gold, marble and mosaics. With a population of more than one million, Rome was the largest city in the ancient world. The streets were crowded with citizens, traders, slaves and visitors from all over the known world. All who saw the city were amazed by its size and magnificence.

The walls of Rome

In 387 BC, in order to protect their city from external threat, the Romans built the so-called Servian Wall around the seven hills. In 271–75 AD, after Rome had expanded beyond this boundary, a second wall, called the Aurelian Wall, was built to enclose the enlarged city.

The city of Rome

Succeeding generations enlarged and improved the city of Rome over many years and in so doing established an influential model for town planners through the centuries.

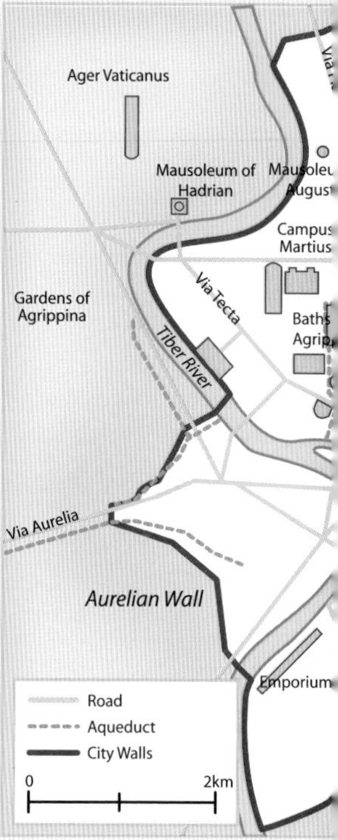

Ager Vaticanus
Mausoleum of Hadrian
Campus Martius
Gardens of Agrippina
Via Tecta
Tiber River
Via Aurelia
Aurelian Wall
Emporium
Road
Aqueduct
City Walls
0
2km

Pincian Hill
Via Pincia
Gardens of Sallust
ardens of omitian
N
Praetorian Camp
Quirinal Hill
Baths of Diocletian
Aurelian Wall
Viminal Hill
Baths of Constantine
apitoline Hill
Baths of Trajan
Esquiline Hill
FORUM
Palatine Hill
Colosseum
Palace ofAugustus
Circus Maximus
Caelian Hill
Via Tusculana
Via Appia
entine Hill
Via Ostiensis
Baths of Caracalla
Via Latina

PART TWO

History of ancient Rome

The history of ancient Rome can be divided into three main periods. The first period, which is shrouded in legend, comprises the reigns of the seven kings of Rome. The second covers the 450 years of the Roman republic, which saw Rome come to dominate most of the known world. The third and final period was that of the Roman empire, in which Roman territory and influence reached its greatest extent.

Imperial Rome
This model shows the layout of Imperial Rome at the time of the emperor Constantine.

THE FOUNDING OF ROME

c. 2000 BC	People from central Europe settle in northwest Italy.
c. 1600 BC	Farmers and shepherds occupy the future site of the city of Rome.
c. 800 BC	Etruscan settlers migrate to western Italy from Asia Minor.
753 BC	Traditional date of the founding of Rome.

The first people to live on the hilltops of what was to become the city of Rome were probably farmers and shepherds of the so-called Latin peoples who had originally migrated to Italy from central Europe. These Latin-speaking settlers established villages and began to trade with neighbouring tribes.

ROMULUS AND REMUS

According to legend, Rome was founded by the twin brothers Romulus and Remus, princes of Alba Longa, a kingdom supposedly founded by the Trojan hero Aeneas after he escaped from Troy. Abandoned as babies on the banks of the Tiber after their throne was seized by their rival Amulius, the infants were suckled by a she-wolf and raised by a shepherd on Rome's Palatine Hill. As adults, they killed Amulius and founded their new city, but quarrelled after Remus mocked his brother by

This celebrated bronze of Romulus and Remus being suckled by a she-wolf dates from around 500 BC.

jumping over the furrow marking the city boundary, provoking Romulus to kill his twin.

The first king of Rome

As the first king of Rome (which was named after him, and is otherwise known as the Eternal City), Romulus populated his new city with wanderers and outlaws. To provide wives, he kidnapped women from the neighbouring Sabine tribe (an incident known as 'The Rape of the Sabine Women'). He is also said to have established a ruling council (Senate) of 100 advisers, and their descendants (called 'patricians') became Rome's leading families.

THE KINGDOM OF ROME

753–509 BC

753 BC	According to legend, Romulus becomes the first of the seven kings of Rome.
c. 750 BC	Greek migrants settle in Sicily and southern Italy.
715 BC	Traditional date of the death of Romulus. He is succeeded by the Sabine king Numa Pompilius.
673 BC	Numa Pompilius is succeeded by the Latin king Tullus Hostilius.
641 BC	Tullus Hostilius is succeeded by the Sabine king Ancus Marcius.
616 BC	Tarquinius Priscus (Tarquin I) becomes the first of three Etruscan kings to rule Rome. During his reign the Forum is laid out.
579 BC	Tarquinius Priscus is succeeded by Servius Tullius.
534 BC	Servius Tullius is succeeded by Tarquinius Superbus (Tarquin the Proud). The first temple on the Capitoline Hill is built.
509 BC	Tarquinius Superbus, the last of the kings of Rome, is overthrown.

The hilltop villages of Rome's early settlers merged into a single city, which grew steadily with waves of new arrivals, including Greeks, Etruscans and Phoenicians from north Africa. By the early sixth century BC, Rome was a substantial city, with temples, a central square called the Forum and defensive walls.

THE SEVEN KINGS OF ROME

According to the historian Livy, Rome was ruled by kings chosen by the Senate, beginning with Romulus and ending with Tarquin the Proud, whose unpopularity led to the establishment of a republic. Tarquin I was the first of three Etruscan kings, who held the first censuses, organized the elections of 300 patricians (nobles) to sit in the Senate and communicated the metalworking and artistic skills of Etruscan craftsmen. Even after the Etruscan kings were expelled, the Etruscan influence remained strong, their varied legacy including the toga and the *fasces* symbol (a hatchet in a bundle of sticks) carried by public officials. The Etruscan kings also built Rome's first drainage system.

A legendary hero

Among the myths of early Rome was that of Horatio Cocles, who single-handedly defended the only bridge into the city from an Etruscan horde while his companions broke it down behind him. According to some versions, he drowned in the Tiber after saving the city; according to others, he swam to safety after the bridge was destroyed.

THE REPUBLIC OF ROME

509–27 BC

509 BC	The republic of Rome is established.
496 BC	After defeat at Lake Regillus, the Romans join an alliance of Latin cities.
494 BC	The plebeians (the urban poor) struggle for power with the patricians (the nobles).
493 BC	The first two tribunes are appointed to defend plebeian interests.
471 BC	A plebeian Assembly is established.
450 BC	The first written set of Roman laws – the Twelve Tables of the Law – is published.
387 BC	Rome is sacked by the Gauls.
367 BC	Under new laws, one consul must be a plebeian.
338 BC	The Romans and the Samnites defeat the other Latin armies.
326 BC	War breaks out between Rome and the Samnites.
312 BC	Work begins on the Appian Way (the first Roman road), and on Rome's first aqueduct.
286 BC	The Romans finally defeat the Samnites.
280 BC	Start of the Pyrrhic wars against King Pyrrhus of Epirus.
275 BC	Pyrrhus is defeated at Beneventum.
c. 264 BC	The Romans control all of Italy.

264–241 BC	The first Punic War with Carthage.
241 BC	Sicily becomes the first Roman province.
238 BC	Sardinia and Corsica fall to Rome.
218–202 BC	The second Punic War with Carthage. Hannibal crosses the Alps, defeating the Romans at Cannae (216 BC), but fails to capture Rome.
204 BC	Rome invades Africa, forcing Hannibal's recall.
202 BC	Scipio defeats Hannibal at Zama.
179 BC	First stone bridge is built over the Tiber.
149–146 BC	The third Punic War ends with the fall of Corinth and Carthage.
90 BC	Italians are granted Roman citizenship.
73–71 BC	The slaves rebel under Spartacus.
60 BC	Crassus, Pompey and Julius Caesar form the First Triumvirate.
58–49 BC	Caesar conducts campaigns in Gaul.
55–54 BC	Caesar invades Britain.
49 BC	Caesar seizes power in Rome, leading to civil war.
48 BC	Pompey is killed at Pharsalus. Caesar meets Cleopatra.
45 BC	Caesar wins the civil war and becomes undisputed leader of Rome.
44 BC	Caesar is elected dictator for life, but is then assassinated.
43 BC	Antony, Octavian and Lepidus form the Second Triumvirate, but civil war restarts.

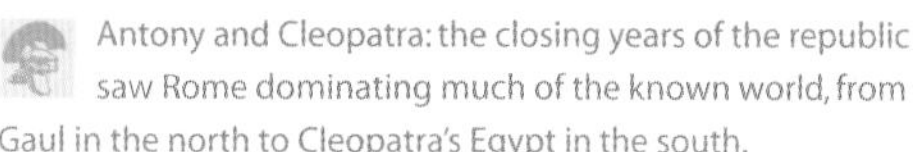

Antony and Cleopatra: the closing years of the republic saw Rome dominating much of the known world, from Gaul in the north to Cleopatra's Egypt in the south.

31 BC Antony and Cleopatra are defeated by Octavian at Actium.

30 BC Antony and Cleopatra commit suicide, and Egypt becomes a Roman possession.

Having thrown out the last of the Etruscan kings, Rome became an independent republic governed by two *praetors* (later renamed consuls), who were elected by the Senate annually. The republic lasted over 450 years and witnessed the expansion of Roman influence over Italy, western Europe, north Africa and the Middle East.

ITALIAN CONQUESTS

Rome grew steadily in power and influence by winning over or conquering hostile neighbouring tribes, such as the Aequi and the Hernici, initially as part of an alliance of Latin cities. Rome eventually came to dominate its allies as well, though it suffered a setback in 387 BC when the city was destroyed by invading Gauls. Conquered enemies became allies of Rome, although they had to supply men to fight in future Roman military campaigns.

In control of Italy, the Romans extended their influence further afield. Carthage, the most powerful of Rome's rivals, was destroyed after a series of wars, leaving Rome the strongest military presence in the known world. Through political treaties and military conquest, Rome

The Capitoline geese

One of the most enduring Roman legends concerned a night attack on the Capitoline Hill during the Gaulish invasion of 387 BC. The defenders were alerted to the threat by the alarmed cackling of the sacred geese kept on the hill and were able to drive the attackers away.

came to control virtually all the countries bordering the Mediterranean, its territories extending to the English Channel in the north, the Spanish peninsular in the west, Egypt in the south and Syria in the east.

LIFE UNDER THE REPUBLIC

A network of roads linked Rome with its remotest possessions, promoting travel and trade and allowing the army to move wherever it was most needed. Rome became the centre of the civilized world. Many Romans made fortunes from trade and enjoyed luxurious lifestyles, keeping slaves to maintain their households.

Although poorer people had fewer privileges, their interests were defended by representatives (tribunes) at the highest levels. Only women and slaves had no real control over their lives. Power remained largely in the hands of the patricians, who dominated the Senate until Julius Caesar ushered in the age of the emperors.

Ideals of the republic

Many other countries devised their own versions of Roman political, legal and social institutions and imitated Roman arts and architecture. The Romans in turn based many of their ideas upon the culture of ancient Greece.

IMPERIAL ROME

27 BC–476 AD

27 BC	Octavian becomes the first Roman emperor, taking the title Augustus.
6 AD	Judaea becomes a Roman province.
14 AD	Augustus dies; succeeded by Tiberius.
37 AD	Tiberius dies; succeeded by Caligula.
41 AD	Caligula assassinated; succeeded by Claudius.
43 AD	The Romans invade Britain.
54 AD	Claudius dies; succeeded by Nero.
64 AD	Rome is badly damaged by fire. Nero orders the persecution of Christians.
68 AD	Nero's suicide is followed by civil war.
69 AD	Galba, Vitellius and Otho are succeeded as emperor by Vespasian.
70 AD	Jerusalem is captured and sacked.
79 AD	Vespasian dies; succeeded by Titus. Pompeii and Herculaneum destroyed by Vesuvius.
80 AD	The Colosseum in Rome opens.
81 AD	Titus dies; succeeded by Domitian.
80 AD	Agricola completes the invasion of Britain, excluding Scotland.
96 AD	Domitian assassinated; succeeded by Nerva.
98 AD	Nerva dies; succeeded by Trajan.
117 AD	Roman empire reaches its largest extent. Trajan dies; succeeded by Hadrian.

The destruction of the Temple in Jerusalem by the army of imperial Rome in 70 AD is still remembered.

122 AD	Construction of Hadrian's Wall begins.
166 AD	Plague rages throughout the empire.
212 AD	Roman citizenship is granted to all free men throughout the empire.
235 AD	Assassination of the emperor Severus Alexander, then fifty years of anarchy under around thirty emperors.
270 AD	Parts of the empire are abandoned.
271–75 AD	Aurelian Wall is built around Rome.
284 AD	The empire is split into East and West.
313 AD	Religious tolerance is introduced.
324 AD	Empire is reunited under Constantine I.
330 AD	Constantine establishes his court at Constantinople.
337 AD	Constantine is baptized a Christian, and dies; empire split between his three sons.
361 AD	State religion is restored under Julian.
367 AD	Barbarian kingdoms are established within Roman territory.

Trajan's empire

The Roman empire reached its height under Trajan in the early second century AD. Rome had defeated countless enemies, including the mighty Persian empire, and had become fabulously rich with plunder from its provinces. The sheer size of this empire, however, was to be its eventual undoing.

394 AD	Empire is reunited under Theodosius I.
402 AD	The Goths invade Italy.
409 AD	The Vandals invade Spain.
410 AD	The last Roman troops leave Britain. Rome is sacked by the Goths.
451 AD	Invasion of Italy by Attila the Hun halted.
455 AD	Rome is sacked by the Vandals.
476 AD	Romulus Augustulus is overthrown and the western empire falls.

By the first century BC the republic had become very corrupt. The authority of the Senate was seriously weakened, especially after Julius Caesar made himself dictator, and, although the Senate still appeared to run public affairs, real power passed to a succession of emperors, starting with Caesar's great-nephew Augustus.

THE PAX ROMANA

With its formidable armies and great wealth, Rome had the status of a world power. Conquered territories were kept firmly under Roman control, and their inhabitants were encouraged to adopt Roman lifestyles. Under the so-called *Pax Romana* (Roman peace), trade prospered and the Roman people were kept happy with lavish public entertainments. The emperors built impressive new temples, palaces and civic buildings, although some behaved so erratically that their reigns ended in assassination or the outbreak of civil war.

The Holy Roman Empire

In 800 AD, Charlemagne, ruler of much of central and western Europe, was proclaimed emperor of the West by Pope Leo III and the heir of the ancient Roman western empire. The last Holy Roman Emperor abdicated in the early nineteenth century.

By the time of Hadrian, the Roman empire was so large that it was virtually ungovernable. Emperors came and went in quick succession, many of them murdered or killed in battle. Rebellions in the provinces, combined with invasion by barbarian armies, economic crises and unrest among the lower classes, led to the gradual abandonment of the empire's distant frontiers and the spread of lawlessness.

CONSTANTINOPLE

In 330 AD, Constantine moved the imperial capital to Constantinople and the empire was split in two, with one emperor in the East and another in the West. Rome was plundered by the Goths in 410 AD, and the last emperor of the West was overthrown in 476 AD. The empire in the East continued as the Byzantine empire, and preserved the laws and culture of Rome for another 1,000 years before eventually falling to the Muslims in 1453 AD.

PART THREE

Roman rulers

The first rulers of ancient Rome were the seven kings, who were overthrown when the republic of Rome was established and replaced by consuls elected by the Senate. The republic collapsed, however, after Julius Caesar became dictator, and was subsequently ruled by a succession of emperors, whose power was absolute.

The first emperor
Augustus, the first of Rome's emperors, is depicted holding court over the arts.

THE SEVEN KINGS OF ROME

According to legend, Rome was founded by Romulus, who ruled as the city's first king. He was succeeded by another six kings, mostly with Sabine or Etruscan backgrounds, whose reigns are a mixture of legend and historical fact. The last king, Tarquin II (otherwise known as Tarquin the Proud) became extremely unpopular and was eventually overthrown, leading to the establishment of a republic.

Although the foundations of later greatness were laid under the kings of Rome, their rule was viewed with great suspicion by later Romans. Many aspiring political leaders (including Julius Caesar) felt obliged to declare publicly that they had no intention of becoming kings themselves.

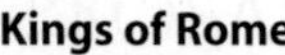

Kings of Rome

753–715 BC	Romulus
715–673 BC	Numa Pompilius
673–641 BC	Tullus Hostilius
641–616 BC	Ancus Marcius
616–579 BC	Lucius Tarquinius Priscus (Tarquin I)
579–534 BC	Servius Tullius
534–509 BC	Lucius Tarquinius Superbus (Tarquin II)

REPUBLICAN LEADERS

Under the republic, power rested ultimately with the Senate and in the two consuls who were elected by them annually. The consuls held power for just one year, although this rule was broken from time to time, and several men were granted the power to rule for a limited time as dictator.

CORRUPTION AND INTRIGUE

The politics of the Roman republic were characterized by corruption and intrigue. In addition to a popular following, successful leaders needed the support of other influential political figures, and many relied on threats, bribes or promises of future favours.

This mural depicts the famous statesman Cicero addressing the Roman senate.

GAIUS MARIUS

ROMAN CONSUL

Born: Arpinum, 157 BC. **Wife**: Julia (aunt of Julius Caesar). **Children**: Gaius Marius the Younger. **Career**: Military tribune, 134 BC; plebeian tribune, 120 BC; consul, 108 BC, 104–100 BC, 86 BC. **Achievements**: Won many military campaigns and reorganized the Roman army; elected consul an unprecedented seven times. **Died**: 86 BC.

Gaius Marius built a reputation as a successful military commander fighting in Spain, Africa and Gaul. Because he was considered the only man who could prevent Germanic tribes invading Italy, he was elected consul in 108 BC, and re-elected every year from 104 to 100 BC. Between 100 BC and 88 BC, he was given command of his rival Sulla's army. The outraged Sulla marched on Rome and Marius was forced into exile. When Sulla left Rome on campaign the following year, Marius returned and was appointed consul for a seventh time.

Military reform

Marius extended military recruitment to citizens who did not own land. As these new recruits were rewarded with gifts of land by their general when they retired, they owed their loyalty to him rather than to the Senate.

LUCIUS CORNELIUS SULLA

ROMAN CONSUL

Born: Rome, 138 BC. **Career**: Pro-consul in Cilicia, 92 BC; consul, 88 BC, 80 BC; dictator, 82–80 BC. **Achievements**: Won notable military victories and, as dictator, tried to restore the republic. **Died**: Puteoli, 78 BC.

Sulla served under Gaius Marius in Africa and warded off Germanic invaders of Italy. After defeating rebellious Italian allies, Sulla was elected consul, triggering a power struggle with Marius. Sulla was about to lead an army against Mithridates, king of Asia Minor, when command was handed to Marius. Sulla ordered his troops into Rome (an illegal act) and Marius was driven into exile. He seized power while Sulla was fighting in Greece and hundreds of Sulla's supporters were executed. After Marius' death, Sulla returned and, as dictator, presided over a reign of terror in which 1500 political opponents were killed. In 80 BC, he handed power back to the Senate and stood for re-election as consul before retiring.

A bad example

Though Sulla tried to ensure that in future no one else should make themselves dictator, his own example undoubtedly contributed to the fall of the republic.

MARCUS LICINIUS CRASSUS

ROMAN CONSUL

Born: *c.* 112 BC. **Career**: Consul, 70 BC, 55 BC. **Achievements**: Defeated slave uprising of 71 BC; formed First Triumvirate with Pompey and Caesar, 60 BC; given control of Syria, 55 BC. **Died**: In battle, 53 BC.

Crassus built a fortune from slave-trafficking and mining. With the aid of Pompey, he suppressed the slave revolt led by Spartacus in 71 BC and had 6000 prisoners crucified. He then demanded that he and Pompey be made consuls. Duly elected, they reversed Sulla's reforms intended to return power to the Senate. In 60 BC, Crassus, Pompey and Julius Caesar formed a secret alliance (the First Triumvirate) and Crassus used his wealth to advance his political ambitions. He served as consul in 55 BC and then was assigned the province of Syria. Seeking glory, he attacked Parthia, only to be captured and executed after defeat at the Battle of Carrhae in 53 BC.

A gruesome death

Legend has it that when Crassus was captured after his final defeat at Carrhae, his captors took exception to his lust for Parthian plunder and killed him by forcing him to swallow molten gold.

POMPEY

ROMAN CONSUL

Born: Picenum, 106 BC. **Career**: Consul, 70 BC, 52 BC. **Achievements**: Won military victories in Africa and Spain; helped suppress slave revolt, 71 BC; cleared the Mediterranean of pirates, 67 BC; captured Jerusalem, 63 BC; formed the First Triumvirate with Crassus and Caesar, 60 BC; waged civil war against Caesar, from 49 BC. **Died**: Assassinated, Egypt, 48 BC.

Gnaeus Pompeius Magnus campaigned under his father (a consul) before siding with Sulla against Marius. He defeated rebels in Sicily, overcame armies in Africa and Spain and completed the rout of the slaves led by Spartacus in 71 BC. His victories in Asia Minor helped to bring the region under Roman control. Pompey and Crassus were elected consuls in 70 BC and set about transferring power from the Senate to themselves. A hero to the common people, Pompey

A political murder

Pompey's murder was ordered by the young pharaoh Ptolemy XIII, who was hoping to win Caesar's favour. When Caesar saw Pompey's severed head, however, he is said to have wept with grief.

formed an alliance (the First Triumvirate) with Crassus and Julius Caesar in 60 BC. He stayed in Rome while Caesar fought in Gaul, but the alliance fell apart. In 52 BC, after the Senate House was burned down by a mob, Pompey was asked to restore order as consul.

CIVIL WAR

In 49 BC Caesar marched on Rome and civil war broke out. Caesar defeated Pompey's supporters in Spain and then Pompey himself at Pharsalus in Greece in 48 BC. Pompey fled to Egypt, where he was murdered.

Julius Caesar displays his grief when presented with the severed head of his rival Pompey.

JULIUS CAESAR

ROMAN DICTATOR

Born: Rome, 100 BC. **Career**: Pontifex Maximus (high priest), 63 BC; consul, 59 BC, 48 BC, 46 BC, 45 BC; proconsul in Gaul, 58–49 BC; dictator, 49 BC; dictator for life, 45 BC. **Achievements**: Formed First Triumvirate with Crassus and Pompey, 60 BC; fought Gallic Wars, 58–49 BC; invaded Britain, 55 BC and 54 BC; won civil war, 49 BC; as dictator introduced reforms and helped transform Rome into an empire. **Died**: Assassinated, Rome, 44 BC.

Gaius Julius Caesar, the most famous Roman of all, was born into a patrician family claiming descent from the goddess Venus. When stripped of his official posts by Sulla, he left Rome to serve in the army. He later sailed to Rhodes to continue his studies but was kidnapped by pirates and ransomed. He achieved public prominence in 63 BC when he won the post of Pontifex Maximus (high priest) and a year later was given control of Spain.

Elected consul in 59 BC, he formed the First Triumvirate with Crassus and Pompey. Between 58 BC and 49 BC, as proconsul for Gaul, he brought virtually all western Europe under Roman control. He led two expeditions to Britain (55 BC and 54 BC). The conquest of Gaul was the greatest military achievement since Alexander the Great and added hugely to Caesar's personal glory. An impressive public speaker, he became hugely popular with his soldiers and the common people of Rome.

CROSSING THE RUBICON

After Pompey was made consul, Caesar was ordered to disband his army. Instead, he broke the law by leading a legion across the Rubicon stream marking the Roman frontier and into Rome. Civil war followed, ending with victory over Pompey at Pharsalus in Greece in 48 BC. Having pursued Pompey to Egypt, Caesar became involved in Egyptian politics and made his lover, Cleopatra VII, queen. A scandal followed when Caesar brought Cleopatra and their son Caesarion to Rome. Declared dictator for life in 45 BC, Caesar introduced some important reforms. He took steps to relieve the hardships of the poor, reduced debts, introduced a new 'Julian' calendar, began major new building projects and reorganized the administration of Rome. He paid little attention to the views of the Senate, however, and some senators, fearing that Caesar had become too powerful, plotted to murder him.

Military genius

Caesar's skill as a military commander was legendary. After one swift victory he is famously reported to have observed: 'Veni, vidi, vici' ('I came, I saw, I conquered').

THE IDES OF MARCH

On the ides of March (15 March) 44 BC, Caesar was assassinated by a group of senators led by Cassius and Caesar's friend Marcus Junius Brutus. According to William Shakespeare, Caesar died with the famous accusation: 'Et tu, Brute? Then fall, Caesar.' The Roman republic effectively died with Caesar. After a period of civil war, the military dictatorship Caesar had created

A soothsayer warns Caesar to 'beware the ides of March' as he passes through the forum before his assassination.

Mark Antony's funeral speech honouring Julius Caesar was the opening shot in a period of political unrest.

continued under his heir Octavian (the first of the Roman emperors) and lasted for another 500 years. Two years after his assassination, Caesar was officially recognized by the Senate as a god.

The Caesars

The name Caesar was borne as a title by all the Roman emperors and later as tsar or czar by the rulers of Russia and as kaiser of Germany. Caesar's first name, meanwhile, is commemorated in the month of July (originally Quintilis).

MARK ANTONY

ROMAN CONSUL

Born: Rome, 83 BC. **Career**: Consul, 44 BC. **Achievements**: Fought in Gaul, 54–50 BC; formed Second Triumvirate with Octavian and Lepidus, 43 BC; met Cleopatra, 41 BC; ruled provinces in the East until defeated by Octavian. **Died**: Committed suicide, Alexandria, Egypt, 30 BC.

Mark Antony was one of Julius Caesar's closest supporters. After service in the army, he became consul in 44 BC. Upon Caesar's death, he formed an alliance with his rivals Octavian (Caesar's heir) and Lepidus and gained control of the Senate. After defeating the armies of Brutus and Cassius, Mark Antony, Octavian and Lepidus divided Roman territory between them, Antony sharing control of Egypt and the East with his lover Cleopatra VII. Antony and Octavian fell out and civil war erupted. After his navy was defeated by Octavian at the Battle of Actium, Antony killed himself by falling on his sword.

Funeral speech

After Caesar's murder in 44 BC, Mark Antony delivered a fine speech in honour of the dead dictator. Seizing the dead man's bloodstained robe, he pointed out the fatal stab wounds, naming each of the murderers in turn.

ROMAN EMPERORS

After the collapse of the Roman republic, supreme power rested for the next 500 years on one man: the emperor. The emperors were not elected – they either inherited the role by virtue of their birth or they were named heir by their predecessor.

NOT ABOVE THE LAW

The emperor was not technically above the law (he wore the toga of ordinary citizens, for instance, and did not wear a crown like the early kings), but many of the emperors lived lives of pleasure and luxury that were very different from those of other Romans. After death, many emperors were worshipped as gods.

THE DECLINE OF IMPERIAL POWER

The lives of the emperors became increasingly perilous as the empire declined from the early second century AD – no less than thirty out of forty-two emperors between 138 AD and 361 AD were murdered or killed in battle. From the third century AD, rival emperors ruled in the East and West. Many of the later emperors were vicious tyrants, whose behaviour led to rebellions and repeated outbreaks of civil war, contributing ultimately to the end of Roman civilization.

Emperors of Rome 27 BC–305 AD

27 BC–14 AD	Augustus
14–37	Tiberius
37–41	Gaius (Caligula)
41–54	Claudius
54–68	Nero
68–69	Galba
69	Otho
69	Vitellius
67–79	Vespasian
79–81	Titus
81–96	Domitian
96–98	Nerva
98–117	Trajan
117–38	Hadrian
138–61	Antoninus Pius
161–80	Marcus Aurelius
161–69	Lucius Verus
180–93	Commodus
193	Pertinax
193	Didius Julianus
193–211	Septimius Severus
211–12	Geta
211–17	Caracalla
217–18	Macrinus
218–22	Elagabalus
222–35	Alexander Severus
235–38	Maximinus
238	Gordian I
238	Gordian II
238	Balbinus
238	Pupienus
238–44	Gordian III
244–49	Philip the Arab
249–51	Decius
251–53	Trebonianus Gallus
251–53	Volusianus
253	Aemilianus
253–60	Valerian
253–68	Gallienus
268–70	Claudius II, Gothicus
270	Quintillus
270–75	Aurelian
275–76	Tacitus
276	Florian
276–82	Probus
282–83	Carus
283–84	Carinus
283–84	Numerianus
284–305	Diocletian

Emperors of Rome 293–476 AD

When Roman territory was split into two empires (East and West), each was ruled by a main emperor (titled Augustus), and his intended heir (titled Caesar).

West

286–305	Maximian (Augustus)
293–305	Constantius Chlorus (Caesar)
305–06	Constantius Chlorus (Augustus)
305	Severus (Caesar)
306–07	Severus (Augustus)
306–12	Maxentius (Augustus)
306–07	Constantine (Caesar)
307–24	Constantine (Augustus)
324–337	Constantine I
317–37	Constantine II (Caesar)
333–37	Constans (Caesar)
337–40	Constantine II (Augustus)
337–50	Constans (Augustus)
350–53	Magnentius (Augustus)
353–61	Constantius II
355–61	Julian (Caesar)
361–63	Julian
363–64	Jovian
364–75	Valentinian I
375–83	Gratian
375–92	Valentinian II
383–88	Magnus Maximus
392–94	Eugenius
392–95	Theodosius I
395–423	Honorius
423–25	Iohannes

425–55	Valentinian III
455	Petronius Maximus
455–456	Avitus
457–461	Majorian
461–465	Libius Severus
467–472	Anthemius
472	Olybrius
473	Glycerius
473–475	Nepos
475–476	Romulus Augustulus

East

286–305	Diocletian (Augustus)
293–305	Galerius (Caesar)
305–11	Galerius (Augustus)
305–09	Maximinus (Caesar)
308–24	Licinius (Augustus)
309–13	Maximinus (Augustus)
317	Licinianus (Caesar)
317–26	Crispus (Caesar)
324–37	Constantius II (Caesar)
335–37	Dalmatius (Caesar)
337–61	Constantius II (Augustus)
350–54	Gallus (Caesar)
364–78	Valens
379–95	Theodosius I
395–408	Arcadius
408–50	Theodosius II
450–57	Marcian
457–474	Leo
474–491	Zeno
475–476	Basiliscus

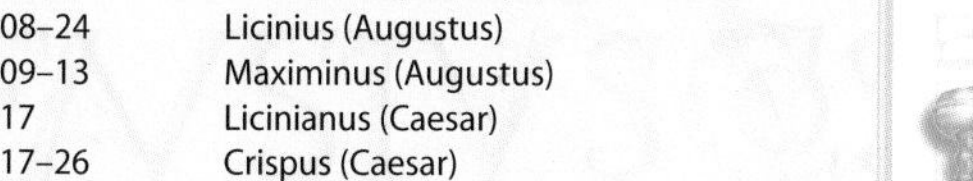

AUGUSTUS

FIRST ROMAN EMPEROR

Born: Rome, 63 BC. **Career**: Formed Second Triumvirate with Mark Antony and Lepidus, 43 BC; defeated Brutus, 42 BC; renamed Augustus on becoming Rome's first emperor, 27 BC. **Achievements**: Avenged Julius Caesar's murder; ended many years of civil war; set pattern for later emperors of Rome. **Died**: Nola, 14 AD.

Born Gaius Julius Caesar Octavianus (Octavian), Augustus was the great-nephew and adopted son of Julius Caesar. After Caesar's assassination, he formed an alliance with Mark Antony and Lepidus and defeated the assassins Cassius and Brutus. The alliance broke down, and in 31 BC he destroyed Antony's fleet at the Battle of Actium. At the Senate's request, Octavian took control of Rome's armies, foreign policy and administration. A brilliant politician, he ostensibly restored Rome's old republic, while keeping power himself, taking responsibility for

Public entertainer

To keep the people entertained, Augustus staged eight major gladiatorial contests (involving 10,000 gladiators), twenty-seven public games and twenty-six wild beast shows (in which some 3500 animals were killed).

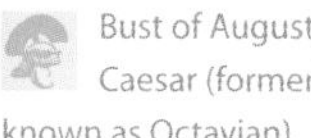
Bust of Augustus Caesar (formerly known as Octavian).

Rome's troubled regions (where the army was). Rome was kept in line by an enlarged Praetorian Guard (his personal bodyguard) and through officials acting as the city's police force.

THE PAX ROMANA

In 27 BC, he accepted the title Augustus ('dignified one') and was declared emperor. He used the wealth of Egypt to build up support within the Senate, which voted according to his wishes. He won popular support by securing Rome's grain supply, generosity to his soldiers and giving money to the common people. He strengthened the army by introducing auxiliary troops (from the provinces) and extended Rome's borders to the Danube and Rhine. This ushered in a new age of peace and prosperity, known as the *Pax Romana*, and Augustus himself was known as the 'second founder of Rome'. At his death in 14 AD, he was declared a god.

TIBERIUS

SECOND ROMAN EMPEROR

Born: 42 BC. **Wives**: Vipsània Agrippina, Julia (daughter of Augustus). **Children**: Julius Caesar Drusus. **Career**: Emperor, 14–37 AD. **Achievements**: Succeeded Augustus as Roman emperor, 14 AD; retired to Capri, 26 AD. **Died**: 37 AD.

The adopted son of Augustus, Tiberius Claudius Nero earned a reputation as a brilliant general, campaigning in Germany and elsewhere before retiring to the island of Rhodes. He was ordered back to Rome in 2 AD and, allegedly against his own wishes, inherited the imperial throne in 14 AD.

Like Augustus before him, Tiberius made a show of allowing the Senate to vote freely, but punished those who criticized him. In most regards he continued the policies of Augustus, and maintained the *Pax Romana* under which the people of Rome prospered. Some of his

The gloomiest of men

A large man with pimples, Tiberius was apparently gloomy and reclusive. He was said to be fond of wine, but to hate crowds and to be scared of thunder.

reforms were unpopular, however, and he lived in terror of being assassinated. He eventually retired to the island of Capri, leaving his deputy Sejanus to act on his behalf. When Sejanus tried to seize power, Tiberius had him executed. From then onwards, anyone Tiberius believed might betray him was forced to commit suicide. It was during the reign of Tiberius that Jesus Christ was crucified by the Romans in Jerusalem.

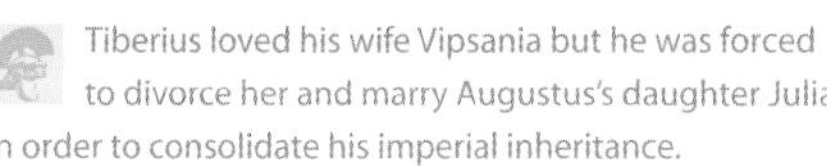

Tiberius loved his wife Vipsania but he was forced to divorce her and marry Augustus's daughter Julia in order to consolidate his imperial inheritance.

CALIGULA

EMPEROR FAMED FOR HIS CRUELTY

Born: Antium, 12 AD. **Wives**: Junia Claudilla, Livia Orestilla, Lollia Paulina, Caesonia. **Children**: Julia Drusilla. **Career**: Emperor, 37–41 AD. **Achievements**: Earned a reputation for extravagance, eccentricity and cruelty. **Died**: Assassinated, 41 AD.

A direct descendant of Augustus, Gaius Julius Caesar Augustus Germanicus was nicknamed Caligula after the *caligulae* (soldiers' boots) that he wore as a child. He was brought up in military camps on the Rhine accompanying his father, a celebrated general, and as an infant was treated as the army's beloved mascot.

As emperor, Caligula proved to be arrogant and cruel, and during the four years that he ruled he became notorious for his extravagance and outlandish, vicious and perverted behaviour. On becoming emperor, he

Bizarre behaviour

Many tales were told of Caligula's madness. As well as declaring himself a living god, he planned to have his favourite horse Incitatus appointed consul and on another occasion ordered his legionaries to collect seashells.

A bust of the cruel Roman emperor Caligula.

initially courted popularity by staging some lavish public entertainments, which included banquets and gladiatorial contests. After a short but serious illness, however, the darker side of Caligula's nature finally revealed itself in outbursts of bizarre and cruel behaviour. He had dozens of people executed for treason and showed little respect for either the Senate or the Roman people.

MADNESS

Caligula's many extraordinary excesses eventually led his contemporaries to conclude that their emperor was not only bad, but also mad. Ultimately, it all became too much, and on 24 January 41 AD, Caligula was murdered by members of his own Praetorian Guard.

CLAUDIUS

FOURTH ROMAN EMPEROR

Born: Lugdunum (Lyons), Gaul, 10 BC. **Career**: Consul, 37 AD; emperor, 41–54 AD. **Achievements**: Conquered Britain, 43 AD; as emperor introduced many reforms. **Died**: Murdered, 54 AD.

Claudius was a grandson of the emperor Tiberius. Although he stammered, dribbled and twitched, he was intelligent, well-read and a fine public speaker. One of the wisest and most popular of all the Roman emperors, he also had a cruel streak according to some historians. After the murder of Caligula, he became emperor reluctantly, on the insistence of the Praetorian Guard. Ruling with an advisory council of freedmen, he presided over trials and attended the Senate, allowing free discussion there. Notable achievements of his reign included the conquest of Britain in 43 AD and allowing citizens from the provinces to become senators.

Poisoned

It is said that Claudius's reign ended prematurely when he ate poisoned mushrooms supplied by his fourth wife, his niece Agrippina. She probably murdered him to ensure that the throne passed to her son Nero.

NERO

ROME'S MAD EMPEROR

Born: Antium, 37 AD. **Career**: Emperor, 54–68 AD. **Achievements**: Peace with Parthia, 63 AD; suppressed British rebellion, 61 AD; designed new city after Great Fire of Rome, 64 AD. **Died**: Committed suicide, 68 AD.

The adopted son of Claudius, and the last of the family of Augustus, Nero inherited the imperial throne at the age of seventeen, at a time when the empire was still prospering under the long peace known as the *Pax Romana*. His reign began well enough, with him heeding the advice of Seneca and other members of the Senate, but subsequently he abandoned himself to drunken revelry, violence and vanity. When his mother Agrippina and other political opponents attempted to intervene, they were murdered. Seneca himself was forced to commit suicide after being implicated in a plot against Nero in 65 AD. It was also said that Nero murdered his pregnant wife Poppaea by kicking her to death.

Nero courted the common people by staging public games and other entertainments, which were paid for by raising taxes. Nero sometimes took part in public entertainments himself, scandalizing Rome by appearing as an actor and singer on stage. Anyone who did not cheer enthusiastically risked being beaten

The excesses of Nero's reign made him one of the most notorious of all the Roman emperors.

by the Praetorian Guard. He went on to win prizes in the Olympic Games for his acting, lute-playing and chariot-racing (even though he fell out of the chariot).

THE GREAT FIRE OF ROME

In 64 AD, Nero was widely suspected of starting the Great Fire of Rome, which destroyed much of the city. Popular – although unlikely – tradition depicts Nero playing his fiddle (or lyre) as the capital burned, happily contemplating the new city that he would then be free to rebuild. He placed the blame for the fire on Rome's Christians and used this as an excuse to order the persecution of the Christian community.

The end came in 68 AD after a serious revolt erupted in Gaul. Declared a public enemy by the Senate and deserted by the Praetorian Guard, Nero was forced to commit suicide. His last words before stabbing himself in the throat with a dagger are said to have been: 'Jupiter, what an artist dies in me!'

The Golden House

Chief among the buildings that Nero proposed for Rome when it was rebuilt after the Great Fire was the palace known as the Golden House. Its name was inspired by the extravagant use of gold leaf throughout the structure.

VESPASIAN

ROMAN SOLDIER-EMPEROR

Born: Falacrina, 9 AD. **Career**: Consul, 51 AD; emperor, 69–79 AD. **Achievements**: Restored order; introduced numerous military and administrative reforms; campaigned successfully in Judaea, 73 AD. **Died**: 79 AD.

Nero's suicide in 68 AD was followed by 'the year of the four emperors', in which the imperial throne was claimed in quick succession by Galba, Otho, Vitellius and, finally, Vespasian. Vespasian was an army commander who had played a major role in the invasion of Britain in 43 AD and served in Africa and Judaea. As emperor, he quickly restored order and reformed the army and administration. He conquered new territory in Germany and Britain and crushed a Jewish rebellion. He embarked on a campaign of public building, including starting work on the Colosseum. The hard-working, plain-thinking, soldier-emperor Vespasian finally died of a fever in 79 AD.

Toilet tax

Vespasian's financial reforms included imposing a tax on the use of public toilets. When others hesitated at handling money from such sources, he mocked them with the words: 'Money does not smell'.

TITUS

GENEROUS AND POPULAR EMPEROR

Born: Rome, 39 AD. **Career**: Military tribune in Britain and Germany, 61–63 AD; emperor, 79–81 AD. **Achievements**: Suppressed the Jewish revolt of 70 AD; completed building of the Colosseum. **Died**: 81 AD.

Titus succeeded his father Vespasian as emperor in 79 AD. He had established a reputation as an army general, helping his father suppress the Jewish revolt in 70 AD by capturing Jerusalem. As emperor, he won popularity through his generous, merciful behaviour. He earned further respect when he provided generous aid to the victims of the eruption of Vesuvius in 79 AD and of a fire that devastated Rome a year later. Other notable achievements included completing the Colosseum and other major buildings. His reign was brief, ending in his death from a fever or, possibly, in his assassination on the orders of his brother Domitian, who succeeded him as emperor. After his death he was declared a god.

Death by mosquito bite

Titus's death may have been caused by a mosquito that flew up his nose. It was claimed his brain was examined and a mosquito the size of a bird was found there.

DOMITIAN

HARSHEST OF THE EMPERORS

Born: Rome, 51 AD. **Career**: Emperor, 81–96 AD. **Achievements**: Strengthened frontier defences against barbarian invasion; gained complete control as censor for life, *c.* 84 AD. **Died**: Assassinated, 96 AD.

Titus Flavius Domitian had to wait many years to become emperor, finally inheriting the imperial throne after the deaths of his father Vespasian and then his brother Titus. The long years of frustrating inactivity, in which he had been entrusted with very little power, had made Domitian impatient and he behaved with great severity once in office, earning a reputation for arrogance and cruelty. He largely ignored the views of the Senate and had senators who dared to question him murdered. His rule became increasingly unforgiving as the years passed, while his acute fear of treachery prompted him to have anyone he suspected of treason put to death. Ultimately, he met a similar fate himself, being murdered in a plot involving his various enemies in the Senate.

Though Domitian had few skills as an administrator, he managed to consolidate the empire's frontiers and also sponsored the arts, constructing new buildings in Rome and restoring others that had been damaged by fire.

NERVA

ENLIGHTENED EMPEROR

Born: *c.* 30 AD. **Career**: Consul, 71 AD, 90 AD; emperor, 96–98 AD. **Achievements**: Restored faith in the imperial system and changed the way in which future emperors were chosen and prepared for office. **Died**: Rome, 98 AD.

The great-grandson of a Roman consul, Nerva was appointed by the Senate to succeed Domitian in 96 AD, by which time he was already an elderly man. Widely respected as a lawyer and twice elected consul, Nerva was one of the most competent emperors. Acting in cooperation with the Senate, he passed laws enabling poorer citizens to hold land of their own, put an end to charges of treason and improved Roman administration. His capable leadership and support for the poor helped restore faith in imperial rule. His choice of Trajan, a celebrated soldier, as his successor similarly met with popular approval from the army.

Choosing an heir

Perhaps the most important of Nerva's innovations was the change he made to the way in which later emperors selected and trained an heir to follow them. This ensured a smooth handover of power and continuity in policy.

TRAJAN

SPANISH-BORN EMPEROR

Born: Italica, Spain, 53 AD. **Career**: Emperor, 98–117 AD. **Achievements**: Introduced important social and administrative reforms; through conquest, extended the Roman empire's borders to their furthest point. **Died**: Cilicia, 117 AD.

Marcus Ulpius Trajan was born in Spain and established a reputation as a capable general, which prompted the emperor Nerva to adopt him as his son and heir in 97 AD. He inherited the imperial throne the following year, becoming the first Roman emperor to come from a provincial family.

Trajan proved a just and popular ruler. As emperor, he introduced many important reforms, which did much to improve the lot of the common people. He secured

Building works

Rome was transformed by Trajan. Among other famous buildings, he ordered the building of the Forum of Trajan, a new aqueduct and a huge bathhouse. Still surviving is Trajan's Column, a massive sculpted pillar bearing 115 scenes recording the emperor's victories in Dacia.

Sculpture of the emperor Trajan.

Rome's corn supply, reduced taxes and put in place an ambitious building programme. Another measure that was established to please the Romans was the holding of 117 consecutive days of public games, in which 11,000 people (mostly Jewish slaves and criminals) were put to death.

TRAJAN'S EMPIRE

The Roman empire reached its greatest extent during the reign of Trajan, through the conquest of Dacia in 105–06 AD, the invasion of Armenia in 114 AD and the conquest of the Persian empire in 115 AD. After conducting a further campaign in Parthia (as a result of which Trajan became the only Roman emperor to reach the Persian Gulf), Trajan died of a stroke in Cilicia on his way back to Rome.

HADRIAN

EMPEROR WHO BUILT HADRIAN'S WALL

Born: Seville, Spain, 76 AD. **Career**: Emperor, 117–38 AD. **Achievements**: Strengthened frontier defences of the Roman empire, 120–131 AD. **Died**: Baiae, 138 AD.

The adopted son of Trajan, Hadrian began his career in the army, fighting in the Parthian campaign. Becoming emperor on Trajan's death, he toured the provinces, improving the empire's frontier defences. A capable military administrator, he built numerous forts, but is best known for Hadrian's Wall, which was constructed across northern Britain to keep at bay the Caledonians north of the border. By the time of Hadrian, the empire had become so large it was virtually ungovernable and, despite his efforts to strengthen the frontiers, his reign marked the beginning of a gradual decline. After 131 AD, he remained in Rome, where he sponsored some major building projects, including the Pantheon and the mausoleum (Castel Sant' Angelo), where he was buried.

Patron of the arts

Hadrian was a scholar and patron of the arts, who revived interest in Greek ideas and encouraged philosophers and writers. He brought back the fashion of wearing beards.

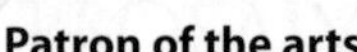

MARCUS AURELIUS

PHILOSOPHER-EMPEROR

Born: 121 AD. **Career**: Co-emperor, 161–169 AD; sole emperor 169–180 AD. **Achievements**: Subdued German tribes; wrote the admired *Meditations*, 170–180 AD. **Died**: Vindobona (Vienna), 180 AD.

At just seventeen, Marcus Aurelius Antoninus Augustus was named co-heir (alongside Lucius Verus) by Hadrian. After serving three terms as consul under the emperor Antoninus Pius, he accordingly became co-emperor with Verus in 161 AD. His reign was marked by almost continual war with various enemies around the empire. Until he died in 169 AD, Verus kept back the Parthians in the East, while Marcus Aurelius fought rebellious tribes in Germany. Marcus Aurelius spent most of his reign on the frontiers of the empire, warding off barbarian invasions; he was fighting on the Danube frontier when he died. For many historians, his death signalled the end of the *Pax Romana*.

An intelligent, hard-working man, Marcus Aurelius is remembered for his philosophical writings, especially his *Meditations*, twelve books of philosophical ideas written whilst campaigning in 170–80 AD. He also passed important social reforms that improved the ordinary lives of slaves and other oppressed classes.

SEPTIMIUS SEVERUS

SOLDIER-EMPEROR

Born: Leptis Magna, Libya, 146 AD. **Career**: Senator, 172 AD; consul, 190 AD; emperor, 193–211 AD. **Achievements**: Restored order after civil war. **Died**: York, 211 AD.

Severus served as a general, senator and consul and as governor of Upper Pannonia before becoming emperor in 193 AD. He spent the early years of his reign fighting and subduing his rivals for the imperial throne. Always a soldier at heart, he expanded the army and ensured his soldiers were better paid. He had little patience with the Senate and had many political opponents executed. He disbanded the Praetorian Guard, replacing it with a force of his loyal supporters. With order restored, he retained popular support by ending the corruption that had characterized the reign of the previous emperor, Commodus. He also promoted the cultural life of Rome. He died while on campaign in Britain, during which he reinforced Hadrian's Wall.

Persecution of Christians

Persecution of the Christians at Lyons and elsewhere continued under Severus, although generally they were shown more tolerance than under other emperors.

DIOCLETIAN

REFORMING EMPEROR

Born: Dioclea, Dalmatia, 245 AD. **Career**: Emperor, 284–305 AD. **Achievements**: Restored order and divided empire into East and West; made important administrative reforms. **Died**: Split, Croatia, 313 AD.

Diocletian rose through the ranks of the army before being proclaimed emperor by his troops in 284 AD. He inherited the imperial throne at a time of chaos and had to defeat his rival Carinus in battle before securing his leadership. To restore order, he set up the so-called Tetrarchy, with the empire being divided into East and West, each with its own emperor. Diocletian, assisted by Galerius, took control of the East, while Maximian ruled the West. Equally significant was his reform of the empire's administrative systems. He made changes to the coinage, tax system and military, as well as ordering the last and most severe persecution of the Christians in 303 AD. In 305 AD, he retired to Salona (Split).

Preserver of the empire

Diocletian's reforms restored both the economy and military strength of the Roman empire, enabling it to remain virtually intact for another 100 years.

CONSTANTINE

FIRST CHRISTIAN EMPEROR

Born: Serbia, 274 AD. **Career**: Emperor 312–37 AD. **Achievements**: Made Christianity state religion, 313 AD; transferred imperial to Constantinople. **Died**: 337 AD.

In 306 AD, Constantine was commanding a Roman army in Britain when his father, the emperor Constantius, died. Constantine was proclaimed emperor by his troops, but the throne was claimed by the tyrannical Maxentius. In 312 AD, before meeting Maxentius in battle, Constantine had a vision of victory under the sign of the Cross. Maxentius died, and Constantine converted to Christianity. Later he sought to reunite the empire.

The emperor Constantine and his rival, the tyrannical Maxentius, meet in battle.

JULIAN

LAST PAGAN EMPEROR

Born: Constantinople, 332 AD. **Career**: Emperor, 360–63 AD. **Achievements**: Military victories in Gaul and Germany; tried to restore pagan religion. **Died**: 363 AD.

Julian, the last surviving relative of Constantine I, was proclaimed emperor by the army in 360 AD, following the death of Constantius II. Julian was the only pagan emperor to occupy the throne after Constantine. Although Christianity was by then the state religion, he attempted to restore Rome's old pagan religion and sought to suppress Christianity by depriving Christian churches of money and decreasing the political influence of Christian bishops. His religious policies made him unpopular, although he also took steps to reduce the size and luxury of the imperial court and made improvements in Rome's civil service. He died while campaigning against the Persians.

Scruffy appearance

Details of Julian's reign have survived in various letters and writings attributed to him. One of his works, entitled *Beard-Hater*, describes how the inhabitants of Antioch mocked him for his beard and scruffy appearance.

PART FOUR

War in ancient Rome

The greatness of Rome was built upon the victories of its legions. The empire was created through military conquest and maintained for hundreds of years by the might of Roman armies, defending Roman territory from both external and internal threats, as well as fighting repeated civil wars. Many of the Roman emperors had backgrounds as soldiers, and spent much of their reigns engaged in military campaigns.

Military strategy

The organization and tactics of Roman armies inspired generations of military commanders.

THE ROLE OF THE ARMY

Rome fought many of its early battles against neighbouring tribes as part of an alliance of Latin cities. Later, however, the Romans went on to establish their own strong, permanent army, with which they gradually conquered all of Italy, and subsequently much of the known world. The conquest of new territory brought back to Rome vast amounts of gold, silver, livestock and slaves. For centuries, the Roman army rarely interfered directly in civilian affairs, and generals gave up their commands on returning to Rome.

By the end of the first century AD, Roman territory had reached its furthest extent, with natural barriers – such as mountains – and the administrative challenges of running such a massive empire hindering any further expansion. The principal role of the legions became defensive, guarding frontiers and quashing rebellions.

Memento mori

An emperor granted a triumph was accompanied in his chariot by a slave, who held a laurel crown above the victor's head. The slave would repeat *'Memento mori'* ('Remember thou art mortal') in the emperor's ear to remind him of the transient nature of his success.

Rome's soldiers built walls and forts of timber or stone on distant frontiers and they manned them for many centuries before being gradually withdrawn to Rome.

TRIUMPHS

Emperors who won great victories with their armies were granted the right to have a triumph, a procession in which the emperor together with his victorious troops marched through Rome with their prisoners and booty. Captured enemy leaders were displayed and, often, put to death in front of huge, cheering crowds.

As well as defeating enemy armies, the Romans were skilled at siege warfare, capturing towns and cities.

THE ROMAN ARMY

The Roman army became the most efficient military force of its time. It was not only one of the largest armies in the ancient world but it was also the best disciplined fighting force.

In Rome's early history, there was no permanent army. In times of trouble, all land-owning citizens who were aged between seventeen and forty-six were summoned to the Campus Martius (Field of Mars) in Rome and there organized into groups of 100 men, which were called 'centuries'. The army was broken up as soon as the danger passed and the soldiers returned to their farms. During the second century BC, however, the need was recognized for a permanent, professional army that was capable of fighting wars at considerable distances from Rome itself.

THE LEGIONS

The backbone of the Roman army was the legion. The number of legions varied over the years. In the second century BC, during the republican era, there were four legions, each comprising around 4200 well-trained and well-armed infantry. The legions were subdivided into maniples of around 120 men. Richer citizens who could afford horses made up the army's cavalry. The number

The standards of the Roman legions were greatly prized and the men who carried them commanded respect.

of legions eventually rose to sixty before being reduced by Emperor Augustus to twenty-eight and subsequently restricted to around thirty.

During the later republican era, the legions comprised centuries of 100 men (later eighty men). The centuries were subdivided into *contubernia* (groups of eight men who shared a tent, a mule and a millstone). Centuries fought together as tactical units which were called cohorts. Each legion had ten cohorts, ranging in size from 480 to 800 men.

RANKS

At the head of each legion was a senior officer called a *legatus*. The silver eagle of the legion was carried by an *aquilifer* (standard-bearer). Less senior standard-bearers often wore an animal skin as a mark of distinction. The cohorts, meanwhile, were each under the command of a *tribunus militum* (military tribune). Each century was commanded by a centurion, whose second-in-command was called an *optio*. The *vexillum*

Losing an eagle

The eagle of a legion was a symbol of the legion's pride. The shame of losing an eagle in battle was so great that the legion that lost it would be disbanded.

(flag) or standard of a century was carried by a *signifer*. Among other ranks was the *cornicen* (who signalled orders by blowing a horn), the *tessarius* (who gave out daily passwords), and the *praefectus castrorum* (who was in charge of building and running camps).

AUXILIARIES

Alongside the legions were various auxiliary troops, which were raised in the provinces. Less well trained than the legions, they made up much of the Roman cavalry (being more used to handling horses than troops from Italy) and filled a range of supporting roles, including the manning of such important frontier fortifications as Hadrian's Wall.

PRAETORIAN GUARD

The emperors maintained an elite unit of several thousand fighting men to protect them; these were known as the Praetorian Guard. These highly-paid soldiers were hand-picked from among the legions and were the only armed forces allowed south of the Rubicon River in northern Italy. First raised by Augustus, the Praetorian Guard tended to serve its own interests before those of the emperors and was often involved in political intrigue, playing a key role in the overthrow or murder of several emperors. It was disbanded by Constantine I in the fourth century AD.

LEGIONARIES

Rome's earliest soldiers were ordinary citizens, who served as required. When the army was reorganized during the later republican period, new laws extended recruitment to those citizens who did not own land. Soldiers received wages, and for the first time Rome had a full-time army. By the second century AD, Rome had 150,000 legionaries (mostly infantry) at its disposal, as well as many more auxiliary troops.

THE LIFE OF A LEGIONARY

Legionaries swore an oath of loyalty to the emperor. They received proper training and were punished for disobedience to their officers. An army career offered opportunities for poorer people and the prospect for a very few of rising through the ranks. Soldiers in the first century AD signed up for 20–25 years. At the end of their service, legionaries were rewarded with gifts of money or land, while auxiliaries were granted Roman citizenship. Against these benefits was the threat of death or serious injury in battle. Soldiers were also forbidden to marry (although many had unofficial wives in the places where they were stationed). The daily lives of Rome's legionaries were hard. When they were not actively campaigning, soldiers might be engaged in building fortifications, roads, bridges and

The soldiers who fought in Rome's legions were the best-disciplined troops in the ancient world.

canals, or they were kept busy drilling or guarding lonely frontiers. Only strict discipline and generous treatment by the emperors prevented mutiny.

WEAPONS AND ARMOUR

Roman legionaries were well-armed and protected. Most infantry fought with a spear or, later, a heavy javelin called a *pilum*, which had a long, narrow metal spike, and could be thrown to break up enemy charges. For fighting at close quarters, legionaries carried a *gladius* (a relatively short-bladed stabbing sword with a grip of wood, bone or ivory), and a *pugio* (a wide-bladed dagger), both copied from Spanish weapons. Auxiliary troops used light throwing javelins and heavier spears as well as bows. For protection, legionaries carried a large rectangular or oval shield made of wood and leather and reinforced with iron.

Body armour consisted of a mail shirt, comprising thousands of bronze scales attached to a fabric shirt, or (from the first century AD) *lorica segmentata* (segmented armour), made of metal strips connected with leather straps and covering the whole of the chest and back. Although heavy, such armour provided excellent protection while still allowing the wearer to move freely. Pieces of leg armour, called greaves, protected the lower leg, while leather strips studded with metal protected the groin.

Under their armour, legionaries wore coarse woollen or linen tunics that extended to mid-thigh, over short breeches or woollen trousers called *bracae*. On their feet they wore *caligulae* (hardwearing sandals).

HELMETS

The first helmets were made of leather or bronze, but these were later replaced by iron helmets with big cheek-plates. Such helmets protected the head, face and neck. Centurions and other officers wore crested helmets that made them conspicuous in battle.

SIEGE MACHINES

The Romans devised special weapons, called *tormenta*, for use when besieging enemy cities. These ranged from siege towers, by means of which Roman soldiers could scale high walls, to battering rams and catapults capable of hurling iron-tipped wooden darts, burning spears and stones at enemy defences, among them the *onager* (a rock-firing catapult) and the *ballista* (a catapult that fired a large iron bolt).

Marching order

Legionaries on the march had to carry all their belongings and equipment with them. Each man was weighed down with weapons, digging tools, armour, a shield, a cloak and a heavy pack containing rations and personal items.

TACTICS

Many of the barbarian armies who fought the Romans relied upon defeating their enemies in one devastating charge, but the Roman army developed much more sophisticated ways of overcoming hostile forces (often much larger than their own). The Roman legionaries were well-disciplined and when attacked they were expected to hold their ground and die at their posts rather than retreat.

FORMATIONS

In the early republican army, maniples of 120 men fought together in a special block formation called a *quincunx*. Experienced soldiers (*triarii*) were positioned at the rear of the quincunx, behind the ranks of heavily-armed ones (*principes*). Younger spear-carrying soldiers (*hastate*) were placed at the front of the formation,

The tortoise

Effective Roman formations included the *testudo* (tortoise), in which a body of men used their shields to create a protective barrier both over their heads and to their front in order to approach the walls of a besieged fortification without taking any casualties.

Individual heroics

Roman legionaries were trained to act as a unit, but individual heroism was also generously rewarded. A man who killed an enemy soldier in single combat was given a drinking bowl, while the first man to scale the wall of a besieged city won a gold crown.

while the poorest and mostly lightly-armed men (*velites*) fought independently of the main formation.

Tight formations

Rome's later legionaries were skilled at fighting in tight formations, interlocking shields to protect themselves from spears and arrows. Infantry usually fought in lines, about 1.2 metres (4 feet) apart. They broke up enemy charges by throwing javelins, relying on their swords and shields at close quarters. As men in the front row fell, the gap was filled by men in the second row.

FIGHTING ELEPHANTS

If attacked by troops with elephants, as used by Hannibal's invading Carthaginian army, the Roman front line would split to let the elephants through and the men behind would then stop them with long spears. The soldiers knew that elephants were difficult to control and that by jabbing them with spears they could be forced to charge back against their own side.

MILITARY CAMPS

When on the march, Roman legions set up substantial camps every night, packing everything up again the next morning. The pitching and striking of these camps was highly organized and followed a set pattern, with rows of tents being arranged behind rapidly-constructed defensive walls of wooden stakes and ditches.

The commander's tent was always located in the middle of the camp, while centurions occupied tents at the end of the row allocated to their century. Other features included hospital tents and watchtowers.

MILITARY FORTS

More permanent forts had stone walls and towers and buildings of wood or stone. Some forts, which housed around 5000 people, became the basis for towns or even cities. Towns, such as York, quickly sprang up beside military forts, their inhabitants attracted by the profits to be made from trading with soldiers. The soldiers themselves often married local women (despite the fact that soldiers were not supposed to take wives while still in service) and thus helped to spread the influence of Rome. Many soldiers settled permanently in these towns with their families after retiring from the army.

Layout of a Roman military camp

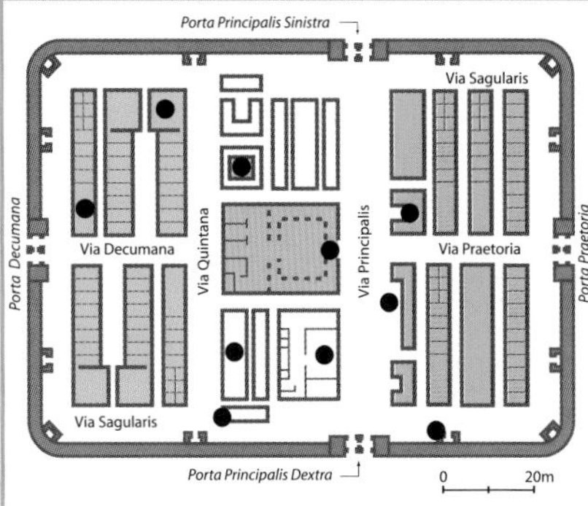

1 *Principia* - Headquarters
2 *Praetorium* - Commander's house
3 *Horrea* - Granaries
4 *Fabrica* - Workshops
5 *Valetudinarium* - Hospital
6 Barracks
7 Stables
8 Houses for Officers
9 Drill Hall
10 Forts

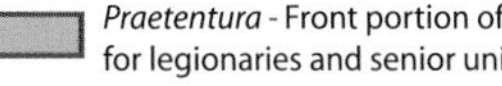

Praetentura - Front portion of the camp for legionaries and senior units

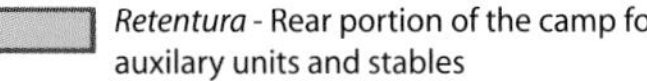

Retentura - Rear portion of the camp for auxilary units and stables

THE ROMAN NAVY

The Romans were not originally a seafaring nation. They constructed their first fleet of warships in 260 BC in response to ancient Carthage's control of the Mediterranean during the first Punic War. This and a second Roman fleet were both lost in storms, but a third fleet finally overcame the Carthaginians in 241 BC, bringing Rome possession of Sicily.

Roman galleys were equipped with boarding planks and grappling hooks to lift enemy vessels out of the water.

In 67 BC, the Roman general Pompey led a fleet of some 200 warships (mostly borrowed from Greece) against the marauding pirate galleys that plagued the Mediterranean and ruthlessly cleared the sea of the pirate menace. A permanent Roman navy was finally established by the emperor Augustus.

WARSHIPS

The earliest Roman warships were all copied from a single captured Carthaginian galley. Most ships were quinqeremes, which were powered by up to five banks of oars and were manned by around 300 oarsmen, or triremes, which had three banks of oars.

Such galleys carried 120 soldiers and were equipped with a ram to sink enemy vessels. A spiked boarding plank called a *corvus* was mounted on the ship's prow and allowed the soldiers to get aboard enemy vessels and capture them. Other weapons included Greek fire (jets of burning liquid) and *ballista* catapults.

War fleets

Roman war galleys sometimes fought together in massed formations of hundreds of vessels. As many as 900 triremes were present at the Battle of Actium between Mark Antony and Octavian in 31 BC.

MAJOR WARS

Roman armies and navies fought numerous wars against their neighbours. Some of these were purely defensive wars in response to invasions of Roman territory. Others were wars of conquest, civil wars or military campaigns fought to suppress internal rebellions. Many of the minor wars are now largely forgotten, but a few of the more important campaigns mark significant milestones in world history.

THE PYRRHIC WARS

In 280 BC, the inhabitants of the Greek city of Tarentum (modern Tarento) in southern Italy objected to Roman interference in their affairs and asked King Pyrrhus of Epirus (in northern Greece) for military aid. Pyrrhus responded by sending 25,000 soldiers. Aided by twenty war elephants (the first seen in Italy), Pyrrhus won battles in 280 BC and 279 BC and penetrated deep into Roman territory, but lost so many men he was driven to admit that 'if we win one more victory against the Romans we shall be totally ruined' (hence the term 'Pyrrhic victory', denoting a victory in which the scale of one's losses outweighs the actual result).

By changing their tactics, the Romans finally defeated Pyrrhus in 275 BC, and Tarentum became an ally of

Rome. This victory marked the first time that Roman armies had defeated those of Greece, and it brought Rome control of the whole of southern Italy.

THE PUNIC WARS

Rome fought three crucial wars with its rival Carthage, the other dominant power in the Mediterranean.

The first Punic War

The first Punic War broke out in 264 BC, and it began well for Rome, with a remarkable naval victory over the more experienced Carthaginian fleet off Mylae in 259 BC. A Roman army sent to Africa under Regulus enjoyed initial success, but it was soundly beaten in 255 BC. A second Roman fleet had to be built after the first was lost in a storm and peace was restored in 241 BC after this replacement fleet managed to defeat the Carthaginians once more, off the Agate Islands.

The second Punic War

The second Punic War began in 218 BC, when the brilliant Carthaginian general Hannibal assembled an army in Spain (then a Carthaginian territory), and struck out for Italy by way of the Alps. In one of the great feats of classical warfare, Hannibal managed to cross the treacherous, snow-laden mountains with most of his army (which included thirty-seven war elephants) intact. He went on to win a series of famous victories against the Roman legions that were sent against him, and, after a crushing defeat at the Battle of Cannae in 216 BC, it seemed the Romans would have to surrender. The Romans, however, built new armies and refused to surrender their capital, even when Hannibal arrived at its walls. Meanwhile, a Roman army under the command of Cornelius Scipio drove the Carthaginians out of Spain (210–206 BC). Denied supplies by the Roman policy of destroying everything in his path, Hannibal returned home to Africa, where he was finally defeated by Scipio at the Battle of Zama in 202 BC. Through this victory, Rome acquired its first overseas provinces.

The third Punic War

In 146 BC, the third Punic War resulted in the final destruction of Carthage. Driven by imperial ambition and by the repeated demand of the great speaker Cato, who ended every speech with the slogan 'Carthage must be destroyed', Roman armies besieged the capital of their long-standing enemy for two years before capturing it and reducing it to rubble, thus effectively bringing to an end Carthaginian power.

THE SLAVE REVOLT

In 73 BC, a widespread revolt among Roman slaves threatened the very existence of Rome itself. Led by a Thracian gladiator named Spartacus, the slaves assembled a huge army and rampaged through

The brilliant and formidable Carthaginian general Hannibal attacked Italy after crossing the Alps.

southern Italy, crushing the legions sent against them. Spartacus then led his fighters northwards towards the Alps, intending to disband his army there, only for his men to refuse to abandon him. After another Roman army was routed by Spartacus and his slaves, the authorities had the survivors decimated (every tenth man being put to death) and then pursued Spartacus southwards towards Sicily, where the slaves hoped to escape Italy on a fleet of pirate ships.

Crassus built a wall across the Isthmus of Bruttium, preventing the slaves escaping, and in 71 BC managed to corner the slave army at Apulia. Surrounded and outnumbered, the slaves were massacred.

Around 7000 captives were subsequently captured by Pompey's troops, after which Pompey himself claimed the credit for putting down the revolt. The captured slaves were crucified beside the Via Appia between Capua and Rome as a warning to others. The body of Spartacus was never identified.

THE CONQUEST OF GAUL

Julius Caesar built his military reputation largely upon his victorious campaigning after becoming proconsul of Gaul in 58 BC. Having defeated the Helvetii in 58 BC, he went on to subdue tribe after tribe, extending the frontier further and further north in Gaul, eventually reaching the Channel and also going into Germany. He then sent military expeditions into Britain in 55 BC and 54 BC and claimed the country for Rome, although it was not until considerably later that permanent Roman control of Britain was established. A serious Gaulish rebellion was suppressed in 52–51 BC, and by the time Caesar returned to Rome in 49 BC, his campaigning in Gaul had cost the lives of around half of all Gauls of military age.

CIVIL WARS

Roman history was plagued by regular outbreaks of civil war, but perhaps the most traumatic of all were those that accompanied the emergence of Julius Caesar as dictator in 49 BC and subsequently broke out again between the various parties claiming power in the wake of Caesar's assassination in 44 BC.

In 49 BC, Caesar followed his rival Pompey to Greece and defeated him at the Battle of Pharsalus in 48 BC. After further successes against the remnants of Pompey's supporters in Africa and Spain, Caesar's supremacy was no longer questioned. Civil war erupted once more, after Caesar's death, between forces loyal to Mark Antony and his lover Cleopatra and those of Caesar's heir Octavian. These ended with the defeat of Antony's navy at the Battle of Actium in northwest Greece in31 BC and Antony's subsequent suicide.

PART FIVE

Religion

The state religion of Rome was based on that of the Greeks, with whom the Romans shared many of the same gods and legends. As devotion to these gods became more formal and less personal, however, many Romans turned to other religions, joining cults dedicated to the worship of foreign gods and goddesses borrowed from Rome's conquered territories, as well as honouring local deities and household guardians. Temples were also built to the emperors of Rome, most of whom were declared to be gods after their deaths.

Gods and goddesses
Some Roman gods and goddesses, such as Venus (pictured here blindfolding her winged son Cupid) are still familiar today long after worship of them died out.

THE STATE RELIGION

The religion of the earliest Romans revolved around the worship of a range of household spirits. It was from this primitive folk religion that the official state religion of Rome developed, much under the influence of the Greeks, who brought their religion with them when they founded colonies in southern Italy. The Romans and the Greeks shared many of the same myths and gods, though they often called them different names.

IMPORTANT GODS

Like the Greeks, the Romans were careful not to offend the gods and tried to win their favour by dedicating shrines and temples to them and sacrificing animals in their honour. They usually depicted the gods in human form and credited them with having human characters and weaknesses. The most important gods, such as Jupiter, Juno, Minerva and Mars, were revered throughout the Roman world, while others were associated with particular places, such as the River Tiber.

LIVING GODS

During the period of the Roman empire, the number of the gods was swelled by the addition of the emperors, most of whom were declared gods after they died, together with their families. Some emperors even went so far as to declare themselves to be gods while they were still alive, although such declarations were not always well-received.

A DOWN-TO-EARTH ATTITUDE

The Romans had an extremely down-to-earth attitude towards the gods and goddesses of the state religion. Most religious ceremonies were carried out with the aim of winning divine favour, usually in the form of practical help of some kind, be it in business, warfare or personal matters. However, the Romans were often equally ready to include in their prayers entreaties to the gods of foreign cults, if they thought that these might be answered.

This made the acceptance of Christianity as the official state religion in the fourth century AD particularly problematic for some, for Christian doctrine forbade the worship of all other gods.

Borrowed deities

The Romans borrowed many of their gods and goddesses from other cultures. These included Juno, Minerva, Venus, Diana and Neptune, borrowed from surrounding Latium, and Ceres from the Etruscans.

GODS AND GODDESSES

Chief among the gods and goddesses of ancient Rome were the twelve Olympians, who ruled in heaven under Jupiter and his queen Juno.

JUPITER

Jupiter was the Roman equivalent of Zeus, the king of the gods to the ancient Greeks. He took his sister Juno as his consort, but was also believed to have had numerous lovers, including other goddesses and mortals and to be the father (by Juno) of Vulcan and (by others) of Diana, Apollo, Minerva, Mercury and Bacchus. He was usually depicted after the Greek fashion as a powerfully-built middle-aged man with a beard, although he lacked the explosive temper of the Greek Zeus.

His symbols were the eagle and the three thunderbolts that he used to strike down those he wished to punish. Jupiter was worshipped as the god of thunder and lightning and the ruler of the earth and sky, controlling the seasons and the passage of the stars through the sky. His chief temple was the Temple of Jupiter on the Capitoline Hill in Rome, the grandest of all the city's temples, where he was worshipped with Juno and Minerva as the Capitoline Triad.

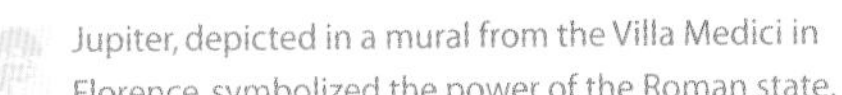

Jupiter, depicted in a mural from the Villa Medici in Florence, symbolized the power of the Roman state.

JUNO

The sister and wife of Jupiter, Juno was the Roman equivalent of Hera, the queen of the gods. She was the mother (by Jupiter) of Vulcan and (through her union with a flower) of Mars. Beautiful and proud, she represented the virtues of Roman motherhood and was revered as the patron goddess of women, marriage, childbirth and children. Unlike the Greeks, the Romans placed less emphasis upon her vengeful anger when betrayed by Jupiter and his mistresses. Juno was one of several gods and goddesses borrowed by the early Romans from the surrounding tribes of Latium. The month of June was named after her. Her symbol was a peacock.

MINERVA

Minerva was the Roman equivalent of Athena, the goddess of wisdom, crafts and war, although she had her roots originally in an Etruscan goddess named Menrfa or Menarva. A daughter of Jupiter by Metis, she was the sister of Diana, Apollo, Mercury, Bacchus and Hercules, among others. She was essentially a warrior-goddess, and was usually depicted wearing armour and a helmet. She was one of the most important of the Roman gods and goddesses, and was worshipped as a patron of doctors and of learning and music. She was also revered as the patron of craftsmen of various kinds, including shipbuilders, carpenters and potters. Her symbol, representing wisdom, was an owl.

MARS

Mars was the Roman equivalent of the Greek Ares, the god of war. A son of Juno, he was a brother of Vulcan and the lover of Venus, among others. He was often

Mars and Minerva both had warring natures and thus were often depicted battling with each other.

depicted as a powerful, bearded soldier in a helmet and armour, more dignified and less short-tempered than his Greek equivalent. Mars was the principal god of Rome's soldiers – generals often made sacrifices to him before going into battle and subsequently presented him with a share of any plunder they captured. Mars gave his name to the month of March and inspired the worship of similar gods under a variety of names in Gaul and Britain. In Rome, his name was given to the

Field of Mars, an area where early Roman armies trained for war. His symbols included twelve sacred spears, which were kept in a shrine on Rome's Palatine Hill.

APOLLO

Apollo (also borrowed from the Greeks) was the Roman god of light. A son of Jupiter and Leto, Apollo was the twin brother of Diana, and was worshipped as the god of the sun, music, poetry, medicine, science, education and prophecy. Roman farmers respected him as a protector of animals, especially sheep and goats. He was usually depicted as a beautiful young man with curly hair, and as a youth was said to have had numerous love affairs with both goddesses and mortals. His symbol was a laurel tree.

DIANA

Diana was the Roman equivalent of the Greek Artemis, the goddess of the moon and of hunting. A daughter of Jupiter by Leto, and the twin sister of Apollo, Diana was revered as the patron of the countryside and wild animals and was celebrated for her skill as a huntress. She was often depicted with a bow and arrow or driving a chariot pulled by stags. Diana never married and had a strong following among women, who revered her as a protector of young girls and pregnant women. Her symbols included cypress trees, deer and dogs.

CERES

Ceres was the Roman equivalent of the Greek Demeter, the goddess of crops and agriculture. A sister of Jupiter, Ceres symbolized the cycle of the seasons and thus new life after death. Her symbols included an ear of corn and a sheaf of barley or wheat.

SATURN

Saturn was the Roman equivalent of the Greek god Cronus, who was revered as a god of agriculture. Saturn was identified as the ruler of Italy in an era before the invention of iron and war. He was the focus of celebration during the annual festival of Saturnalia. The public treasury and (in peacetime) the standards of the legions were kept in his temple in the Forum in Rome. Saturday was named in his honour.

MERCURY

Mercury was the Roman equivalent of the Greek Hermes, the messenger of the gods and the god of trade and thieves. A son of Jupiter, Mercury was believed to carry messages from the gods to the mortal world and was accordingly also worshipped as the patron of travellers, often being depicted wearing a traveller's hat and winged sandals. His symbol was the caduceus, a staff entwined with two serpents.

VENUS

Venus was the Roman equivalent of the Greek Aphrodite, the goddess of love and beauty (although she probably began as a minor nature goddess associated with spring before the link was made with Greek mythology). A daughter of Jupiter, she was identified as the wife of Vulcan, but had many other lovers, including Mars and Mercury. Her children were said to include Cupid (famous for firing arrows that inspired love in those they hit) and the hero Aeneas, sometimes considered the founder of the Roman race.

She was depicted as a beautiful young woman, usually naked or semi-naked, and according to one legend was born from the sea, floating to shore on a scallop shell. Her beauty was the cause of much unrest among the gods; it was supposedly due to her disruptive influence that the Trojan War broke out, after the jealous Juno and Minerva challenged the Trojan prince Paris to decide whether he found either of them or Venus the most beautiful.

An ancestor of Julius Caesar

Venus was claimed as an ancestor of the family of Julius Caesar, and in 46 BC he built a temple in her honour in Rome's Forum. Her name is still familiar today as that of the evening star, the planet Venus. Her symbols included roses, doves, sparrows, dolphins and rams.

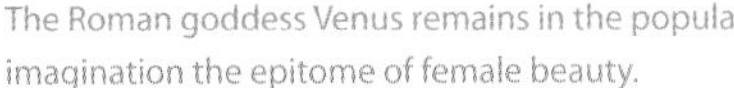

The Roman goddess Venus remains in the popular imagination the epitome of female beauty.

VULCAN

Vulcan was the Roman blacksmith god, equivalent to the Greek Hephaestos, god of fire and metalworking and the patron of craftsmen. He was a son of Jupiter and Juno, and, despite his great ugliness, became the husband of Venus, the goddess of beauty and love. Venus was unfaithful and he took revenge by trapping her and her lover Mars in a fine metal net as they slept. He made many magical objects for the gods with his mighty blacksmith's hammer, among them armour that made the wearer invincible and a golden throne for Juno. As Volcanus, he was one of the gods borrowed by the early Romans from the surrounding tribes of Latium, and was revered as a protector of Rome itself.

VESTA

Vesta was the Roman equivalent of the Greek Hestia, the goddess of the hearth and home. A daughter of Cronus and Rhea, Vesta refused marriage to Neptune and Apollo and remained chaste. She was worshipped as the patron of the family, and newborn babies were presented to her at the hearth. The temple of Vesta in the Forum in Rome housed an Eternal Flame, which was constantly attended by six Vestal Virgins. The Vestal Virgins were chosen by lot from Rome's leading families at the age of six, and were greatly respected. They served in the role for thirty years, during which time they were not allowed to marry or break their vow of chastity. Vestal Virgins who failed to keep their vow could be whipped to death or buried alive.

NEPTUNE

Neptune was the Roman equivalent of the Greek Poseidon, the god of the sea. A brother of Jupiter, he was said to live in an underwater palace and to roam the oceans in a gold chariot pulled by white horses. He could cause the seas to be calm or rough, using his three-pronged trident to whip up storms or tidal waves. He was usually depicted as a strong middle-aged man with long hair and a beard. Travellers at sea commonly prayed to him to keep them safe on their journeys.

JANUS

Janus was one of the few Roman gods who had no direct Greek equivalent. Depicted with two faces, he was the god of doorways and gates, and thus also the god of new beginnings (hence the naming of January, the first month of the new year, in his honour). His popular cult following was established early in Rome's history, supposedly by Romulus. His emblems included keys and the stick used by the guardians of doors and gates to drive away undesirable strangers. Even today a person who is suspected of having divided loyalties may be described as being 'Janus-faced'.

CULTS

By the time of Augustus, most educated Romans no longer believed in the gods of the old formal state religion and sought more personal creeds. Large numbers of Romans joined religious cults, many of which were imported from conquered territories. While the cult of the emperors was actively promoted by the ruling class as a means of strengthening the unity of the empire, some of these alternative religions were regarded with suspicion and were actually banned by the authorities from time to time.

CULT OF THE EMPERORS

The Romans inherited from the Greeks and other eastern Mediterranean countries the custom of worshipping their rulers as gods. The first emperor Augustus, like many of his successors, was careful not to proclaim his own divinity while alive, knowing that such a move would anger fellow-Romans, but he did, however, allow altars and temples to be dedicated to him in the provinces (where emperor worship was always at its strongest) and was declared a god after his death, just as Julius Caesar had been before him. Most of the other emperors were similarly declared gods once dead, although some of the more notorious rulers, such as Tiberius, Caligula and Nero, were refused this honour by the Senate. Some emperors, such as Caligula and Nero, took the idea of their divinity very seriously, and insisted upon being treated as gods while still alive. The emperors were often worshipped in association with Roma Dea, the goddess of Rome itself.

CULT OF BACCHUS

The cult of Bacchus ranked among the more notorious of Roman cults. Bacchus was the Roman equivalent of the Greek Dionysus, the god of wine, revelry and fertility. A son of Jupiter, he was often depicted as a jolly, bearded fat man with a huge belly and ivy in his hair – though he might also appear in the form of a bull or goat or as a beautiful youth.

The secretive celebrations known as the Bacchanalia, held each March in honour of Bacchus, were frenzied, with much drinking and dancing by his followers (called Bacchantes). People allowed into the cult had to undergo a mysterious initiation ceremony, which appears to have involved music and the whipping of new recruits (the majority of whom were female). In 186 BC, the Senate, fearing that public disorder would result from the festivities pursued by the followers of Bacchus, prohibited worship of the god, forcing his followers to continue their rites in private for some years until the law was eventually relaxed. The god's symbols included a bunch of grapes and a staff.

CULT OF CYBELE

The cult of Cybele was a particularly controversial fertility cult. Followers of the goddess Cybele (a nature and fertility goddess also known as the Great Mother), whipped and castrated themselves in the course of blood-soaked rituals in order to become priests of the goddess. A prophecy that Rome would only be saved from Hannibal and the Carthaginians if it embraced the cult of Cybele persuaded many Romans to take up her worship. Despite laws forbidding ordinary Romans from acting as priests, and despite the barbaric nature of their rites, the priests of Cybele were allowed to conduct ceremonies in their own temple on the Palatine Hill, and under Claudius early in the first century AD, the cult of Cybele was formally made part of the official state religion.

CULT OF PYTHAGORAS

Some Romans were attracted to the philosophical theories of the Greek mathematician Pythagoras, in particular to his notion that the souls of mortal men were reincarnated in animals or plants depending upon how they had led their lives. This mysterious cult (which recommended vegetarianism among other ideas) was taken up by Roman intellectuals in the first century BC, although its followers were often accused of practising magic and subjected to official repression.

CULT OF ISIS

The Romans, like the Greeks, were largely unimpressed by most of the gods worshipped by the Egyptians, but Isis was an exception. The sister and wife of Osiris, Isis was a fertility goddess whose worship spread throughout the Mediterranean, and was promoted by Cleopatra when she came to Rome in 45 BC. Celebrated at a big festival each spring, Isis had a strong following among women, freedmen and slaves, although some of the emperors considered her worship undesirable. Under Caligula, however, the cult enjoyed imperial approval, and temples were dedicated to her in the Field of Mars in Rome, at Pompeii and elsewhere. The cult of Isis became increasingly widespread during the third century AD and after its decline in the sixth century AD, some of the goddess's attributes (such as her blue robe) were absorbed into the cult of the Virgin Mary.

CULT OF SERAPIS

Another deity of Egyptian origin adopted by the Romans, Serapis resulted originally through the combination of aspects of the god Osiris and the sacred Apis bulls, and was later associated with Jupiter. Vespasian assumed the title of Serapis on claiming the imperial throne in 69 AD, and in the third century AD Caracalla built a temple of Serapis on Rome's Quirinal Hill. Followers of Serapis valued the god's reputation as a healing god.

CULT OF EPONA

The cult of Epona involved worship of a Celtic goddess of horses called Epona. It became popular among Roman soldiers serving in Gaul and was probably taken by them to other parts of the empire, including Spain, Germany and Scotland, as well as to Italy.

CULT OF MITHRAS

Mithras was a sun god worshipped in India and Persia, who was adopted by the Romans in the first century BC. Most cult members were soldiers, and they were ranked in much the same fashion as in the army. Because of its links with the military, the cult spread wherever the army went and had followers throughout the empire. Notable members included the emperor Nero, who once dressed as Mithras in an official ceremony, and the emperors Hadrian, Commodus and (allegedly) Julian. In certain respects, in particular the bonds of loyalty between worshippers and the promise of an afterlife, the cult of Mithras shared much in common with the emerging cult of Christianity. The cult finally died out in the fifth century AD.

CULT OF SOL INVICTUS

Associated with the cult of Mithras was that of Sol Invictus (the Unconquered Sun). This cult emerged in

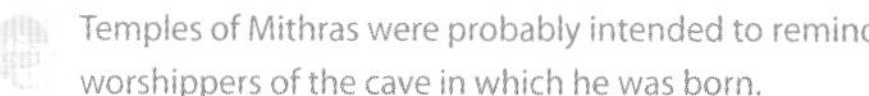
Temples of Mithras were probably intended to remind worshippers of the cave in which he was born.

the second century AD when attempts were being made to bring all Rome's many gods together. Worship of the sun peaked under the emperor Elagabalus, a follower of the Syrian sun god Baal, but the extreme nature of the rites associated with Baal (including human sacrifice) repelled most Romans, and it was not until the third century AD that the emperor Aurelian created an official cult of the sun, believing this would help unify the disintegrating empire. Constantine I, the first Christian emperor, was a follower of Sol Invictus in his early years. The feast day of Sol Invictus, 25 December, was chosen by the Christian church for Christmas.

HOUSEHOLD GODS

As well as the gods of the official state religion and those of various cults, the Romans worshipped *numina* (household gods), most of whom fell into two groups, known as the lares and the *penates*. It was believed that every house had one or more guardian spirits, known as its *lares*, whose protection would be requested at important times in the household, such as weddings and funerals. Two further spirits, called the penates, presided over the family's supply of food and drink.

HOUSEHOLD SHRINES

Most houses contained a *lararium* (a shrine), where people prayed and made daily offerings to these family guardians, and also to the *genius* (the guardian spirit of the family) and to the *manes* (ancestors). The household gods themselves were often depicted in the form of a snake.

Golden figurines

Household shrines of rich families might contain valuable gold figurines of family guardians. The emperor Alexander Severus, for instance, was said to worship at two larariums containing statues of Christ, Abraham, Orpheus, Alexander the Great, Virgil, Cicero and Achilles among others.

JUDAISM AND CHRISTIANITY

The Christian religion was founded in Palestine in the first century AD, at a time when the region was under Roman occupation. The Romans came into contact with Judaism after conquering Palestine in 63 BC. Although tolerant initially, they later persecuted the Jews as their worship of one god prevented them from acknowledging the divine status of the Roman emperors (something the Romans considered a threat to the unity of the empire). There were numerous Jewish

Many Christians were thrown before wild beasts in the Circus Maximus to entertain the ordinary Romans.

revolts against the Romans, which were suppressed with great severity. In 70 AD, the Romans destroyed the main Jewish temple in Jerusalem.

PERSECUTION OF CHRISTIANS

The new Christian religion gathered pace after the crucifixion of Jesus Christ around 33 AD. The refusal of Christians to worship Roman emperors or to make sacrifices to Roman gods led to mass persecution of Christians. Nero, for instance, accused them of starting the Great Fire of Rome in 64 AD and had large numbers of Christian captives publicly burned, crucified or torn apart by wild beasts. Christianity offered comfort to the poor and oppressed and the prospect of an afterlife, thus attracting increasing numbers of Romans, who met for secret services in the underground catacombs of Rome and elsewhere throughout the empire. Persecution reached a peak in the third century AD, but the situation improved under Constantine I (the first Christian emperor), from 313 AD, and in 394 AD Christianity became the official state religion.

Home of the popes

After the martyrdom of Saint Peter (founder of the Roman Catholic church) in 67 AD, Rome emerged as an important Christian centre and home to the Roman Catholic popes.

TEMPLES

All the major Roman towns and cities boasted large temples, most of which were dedicated to a particular god or goddess. The temple was considered the god's home and usually contained a statue of the god concerned. Religious ceremonies took place outside the temple, at an altar placed in a courtyard before the building. The temple might also be used to store plunder seized in the course of military campaigns as well as items of value donated by citizens.

The Pantheon in Rome is among the best preserved examples of a Roman temple anywhere in the world.

LAYOUT

The layout of many Roman temples imitated that of Etruscan and Greek temples, having a rectangular shape and rows of pillars supporting a triangular portico. Some, however, varied in design, with circular and even triangular layouts. The Pantheon in Rome, which still stands, was remarkable for its massive circular dome of concrete, which opened to the sky at its centre. Unlike Greek temples, Roman temples were placed on a raised stone platform to make them even more impressive to worshippers.

The Temple of Jupiter in Rome (now lost) was the most important temple. Several temples have survived in a more or less intact condition, among them the Temple of Portunus in Rome, the Temple of Bacchus at Baalbek in Lebanon, the Temple of Rome and Augustus at Pula in Croatia, the Temple of Augustus and Livia at Vienne and the Maison Carrée at Nîmes (both in France).

Attendance

Most Romans visited temples only occasionally, usually to take part in festivals or other ceremonial occasions. When in need of a particular divine favour, they visited the temple of the most appropriate deity (for instance, Jupiter for victory in battle, or Venus for luck in love).

RITUALS

The rituals of the official state religion were extremely formal and ceremonial in nature, with the ordinary Roman people taking relatively little part in them. It was thought important to retain divine favour through sacrifices and elaborate public ceremonies, which had to be carried out correctly if they were to please the gods. While one priest conducted the ritual, another checked to make sure that it was done correctly while yet another played a flute to drown out distracting outside noises.

FAVOURS AND OFFERINGS

Worshippers asked for favours or offered thanks for favours received. They also invoked the gods when making important vows. Offerings to the gods from wealthy citizens ranged from food, drink and incense (to be burned in sacrificial fires) to coins, brooches, silver statues and even whole temples. Poorer people offered flowers or cakes.

SACRIFICES

Animal sacrifices were very common. The animals presented at the altar ranged from doves, sheep, pigs and goats to oxen and whole herds of cattle. Under the

Influencing the gods

The power of sacrifice to influence the gods was taken extremely seriously. Roman soldiers might even make sacrifices to the gods of their enemies in the belief that this might win them over.

strict supervision of the priests, these were sprinkled with salt, flour or wine and then killed by having their throats cut or, in the case of larger animals, being felled with an axe. The best parts of the carcass were placed in a sacred fire for the gods to consume. The entrails could then be inspected for any omens they might reveal about the future. Any meat left over was eaten as a sacrificial meal.

KEEPING THE GODS HAPPY

With so many gods and goddesses to keep happy, the Romans worried that some might be offended if they felt they had been forgotten during religious ceremonies, and they might be tempted to revenge themselves upon those present by bringing them misfortune or ignoring their requests. To counter this, the priests often addressed their rituals to 'the gods who inhabit this place' to make sure that no god or goddess would feel left out and that the *pax deorum* (goodwill of the gods) was not disrupted.

PRIESTS

Rituals in Roman temples were presided over by priests, who also looked after the statues of the gods and goddesses. Assisted by their attendants, they also conducted sacrifices and took a prominent role in public festivals. Roman priests usually had other duties in public life, and (with the exception of the Vestal Virgins) assumed their priestly role only as needed, pulling their togas up over their heads to indicate that they were now acting as priests.

HIGH PRIESTS

The priests of Rome enjoyed considerable respect as important state officials and were selected from the city's leading families. Julius Caesar as a young man fulfilled the role of head priest, and subsequently the emperor Augustus also made himself head priest, assuming the title Pontifex Maximus ('highest priest'). The title was later passed on to other emperors and ultimately to the popes of the Roman Catholic church.

Readers of omens

Some priests were specially trained in the art of prophecy and were often consulted to read the meaning of omens.

OMENS AND AUGURIES

Most Romans believed that it was possible to predict the future by magical means. Priests called augurs predicted the wishes of the gods through observation of the behaviour of birds, the shapes of clouds, lightning and other natural phenomena. Other priests, known as *haruspices*, examined the entrails of sacrificed animals (paying particular attention to the shape and condition of the liver) for clues about future events. This was a custom inherited from the Etruscans.

Great faith was placed in the writings of the Cumaean Sibyl, a prophetess of the sixth century BC.

Many Romans refused to travel, get married or fight battles if omens suggested it was inadvisable. They also believed that certain days were unlucky. The Senate never met on these *dies nefasti* and little business was done. Romans considered the left (sinister) side to be unlucky – Augustus was among many Roman citizens who always put their right shoe on before their left shoe to avoid bad luck. The Romans also consulted astrologers, who made predictions based on the positions of the stars. Other methods of reading the future included palm-reading and the throwing of dice.

THE SIBYLLINE BOOKS

In times of national crisis, the Romans consulted the Sibylline books, which were stored in the Capitol in Rome. These nine volumes of prophecy were supposedly written by the Sibyl of Cumae (near Naples), a priestess of Apollo who sold them to King Tarquin in the sixth century BC. They were destroyed during the invasions of the fifth century AD.

Curses

In order to get revenge on their enemies, Romans scratched curses on lead tablets and then left them in a temple or some other sacred place in the belief that these would magically harm the person named.

FESTIVALS

The Roman calendar included numerous festivals, which were often marked with processions, feasting, music, dancing, theatrical performances and various sporting contests as well as sacrifices and other religious ceremonies. During the reign of Augustus there were at least 155 festivals annually, though this figure later rose to over 200. Each festival had its own meaning and character, some lasting a single day and others going on for as long as a week. Many were public holidays, although business did not cease altogether except on the most significant occasions.

MAJOR HOLIDAYS

Most important of all the Roman festivals was that of Saturnalia, which began on 17 December and lasted several days. Held in honour of Saturn, it was celebrated

Christian festivals

After Christianity became the official state religion in the fourth century AD the festivals continued, although now combined with Christian feasts. Lupercalia became St Valentine's Day, while Saturnalia was absorbed into the Christian church's Christmas celebrations.

with much feasting, drinking and the giving of presents. Several festivals were closely linked to agriculture, among them Compitalia (held in early January), in which farms were purified for the coming year, the Ludi Ceriales (12–19 April), in honour of the corn goddess Ceres, and Parilia (21 April), which began as a country festival intended to win divine protection for livestock. Other festivals, such as Parentalia (13–21 February), held in honour of dead parents, and Caristia (22 February), celebrated the family.

SPORTS AND CELEBRATIONS

Many festivals became little more than good excuses for sporting and other activities. During the festival of Lupercalia (15 February), for instance, young men covered in blood ran around the Palatine Hill, whipping spectators with strips of goatskin and thus promoting their fertility. Other festivals, such as the festival of Mars (14 March), the Ludi Apollinares (6–13 July) and the Ludi Plebeii (4–7 November), were celebrated with horse racing or gladiatorial contests.

The gods themselves were invited to enjoy the celebrations that took place during the Ludi Romani (5–19 September), with statues of Jupiter, Juno and Minerva being dressed and laid on couches so that they could participate in the revels, which included drama and chariot racing.

PART SIX

Roman society

Ancient Roman society was based on a fairly rigid class system, with citizens enjoying far greater rights than others, especially slaves. Through a combination of personal accounts and archaeological remains we now have a fairly rich and detailed knowledge of how the ancient Romans lived. Through conquest, many features of the ancient Roman way of life came to be copied throughout the empire and their legacy remains strong in many modern cultures.

A complex society

Roman emperors presided over a complex society, with many social ranks and a very detailed system of laws and government.

SOCIAL CLASSES

After the expulsion of its last king in 509 BC, Rome became a republic. People of Roman origin enjoyed full rights as citizens (*cives*), while others were classed as foreigners (*peregrini*). As Rome expanded under the empire, its population mostly comprised citizens, provincials, freedmen and slaves. Although it was sometimes possible to move from one class to another, Rome's class system was fairly rigid and often decided the course of a person's life.

CITIZENS

Citizens were allowed to vote in elections and were originally the only people permitted to join the army. Under the early republic, only people with Roman parents qualified as citizens, although from the first century AD many people from the provinces were also granted citizenship, and army recruits were drawn from wider sources. The emperors themselves were technically citizens, and thus subject to the laws of the state, although many came from a military background or considered themselves superior to the broad mass of the populace and even to have semi-divine status.

Classes of citizens

There were three classes of Roman citizen. The leading

noble families and rich landowners belonged to the influential patrician class. People holding important administrative posts were termed *equites* or equestrians (being mostly descended from Roman cavalry officers). Below them came the mass of the common people,who were known as plebeians. Many plebeians were very poor, but they were still entitled to vote in elections.

PROVINCIALS

People from Rome's provinces did not enjoy the rights of full citizens. They had no vote in elections, and, unlike citizens, had to pay taxes. Some provinces resisted Roman domination, but most quickly saw the benefits to be gained from cooperation with such powerful invaders and they soon adopted Roman ways.

The spread of Roman culture

People throughout the provinces learned Latin, wore Roman togas and imitated Roman manners. Their architects copied Roman villas and temples, while ordinary people relaxed at public bathhouses or amphitheatres much as the Romans themselves did. Some provincials were allowed to purchase Roman citizenship and to serve in Rome's armies. A select few even rose to the rank of senator and attended the Senate in Rome. Indeed, some of the later emperors of Rome hailed from the provinces.

FREEDMEN

Freedmen were former slaves who had been granted their freedom by their masters, usually only after years of long and faithful service. One option for a slave who received wages was to save any money he was given by his master until he had enough to buy his freedom, which was awarded in a special ceremony.

By the time of the Roman empire, freed slaves made up a significant part of the population, and enjoyed considerable influence, particularly in business and administration.

SLAVES

Slaves occupied the lowest rung of society and had no rights at all, being bought and sold like any other kind of property (a good slave typically cost upwards of 2000 denarii). A man's status was often judged by the number of slaves he owned, and Rome's slave

Women

Women were not considered citizens and had few legal rights, depending instead upon their fathers or husbands. In many respects, they were little better off than the slaves who worked in their households.

population numbered many thousands by the time of the late republic. Most slaves were prisoners of war, criminals or themselves the children of slaves.

Everyday life and work

A slave's life was in the hands of his master: some were treated badly, while others were treated well and even rose to positions of power (Rome's civil service, for instance, was largely run by freedmen and slaves). The Roman economy depended largely upon slave labour. Many slaves were employed by the state as builders or other kinds of manual workers. A few were selected to fight as gladiators in the arena or to race chariots during public games. Others were taken on by families to work as household servants or as farmhands. Clever slaves, especially Greeks, worked as doctors or tutors, among other roles. The harshest conditions of all were endured by the slaves who worked in factories and in the mines – their lives were hard and the work was often dangerous.

Punishments

Slaves were expected to behave themselves and punishments were brutal if they broke the rules. For instance, if a slave killed his master, all the slaves in the household were put to death. Such severity provoked several slave rebellions, notably the slave revolt of 73 BC led by Spartacus, which led to the crucifixion of thousands of his followers.

GOVERNMENT

Rome was never a democracy in the way that some of the Greek city-states were. Under the republic that was established in 509 BC, Rome was ruled by three main institutions, the people (as represented by the *comitia*, or Assembly), the Senate (effectively, a council of elders) and a body of elected magistrates who were elected by the *comitia*. The Assembly, however, met rarely and only on the summons of the magistrates, and real power rested with the Senate.

THE SENATE

The Senate originally comprised 100 senators (all of them former magistrates), although by 82 BC their number had risen to 600. Senators, who mostly came from the patrician class of rich landowners, were unpaid. The Senate made decisions on behalf of the

SPQR

The notion that the Senate acted on behalf of the people was summarized in the initials SPQR (an abbreviation of *Senatus Populusque Romanus*, meaning 'the Senate and the People of Rome'). These letters were inscribed on coins, official documents and the standards of the Roman legions.

people and provided a forum for the discussion of political issues. Statesmen also made speeches outside the Senate House (the Curia) in the Forum, addressing crowds from a platform that was built from the prows of captured enemy ships. Some senators, such as Cicero (106–43 BC), were respected by their contemporaries, and are still remembered for their great skill as political leaders and public speakers, although greed and corruption were recurring themes in the Senate's history.

Dictators and leadership

In times of crisis, the Senate might appoint a dictator, who enjoyed absolute authority, assisted by a *magister equitum* (master of the cavalry). Such powers were granted for a temporary period, usually for no more than six months. Julius Caesar challenged the authority of an already-weakened Senate when he became dictator for life, and when the republic gave way to the empire under Augustus and his successors, the Senate was increasingly ignored and reduced to merely ratifying imperial decisions. Augustus made a show of offering power back to the Senate, knowing that Rome needed his leadership to maintain order.

The office of senator later became largely hereditary, being retained within families, although the post was opened up to men from elsewhere in the empire. Rome's Senate House, which was rebuilt in the Forum in the fourth century AD, still stands.

THE MAGISTRATES

The most senior magistrates elected by the Assembly were the two consuls, who shared responsibility for the city's affairs and also served as generals if war broke out. After their year in office they became life members of the Senate, and might be appointed proconsuls (governors of provinces). Below them were a number of other official posts, including eight (later sixteen) *praetores* (judges), four *aediles* (supervisors of markets, streets, public buildings and organizers of public games), two *censores* (who admitted and removed existing senators and oversaw state contracts and tax collection) and twenty *quaestores* (financial administrators).

THE ASSEMBLY

The Assembly was founded in the era of the kingdom of Rome, when its main role lay in organizing the population as a military force. Initially it was dominated by Rome's richer citizens. This situation led to unrest among discontented plebeians, who staged strikes

Symbols of office

Important officials were escorted in public by lictors, who carried the *fasces* (an axe in a bundle of rods tied with a strap) as a symbol of their authority.

Cicero addressing Catiline during a session of the Roman Senate.

until it was agreed (in 450 BC) that the laws of Rome would be properly written down, thereby limiting the power of the magistrates to act as they liked. The plebeians were also granted their own officials, called tribunes, to represent their interests (though few did this very enthusiastically).

Citizens attending the Assembly voted to accept or reject the new laws that were proposed by the Senate.

In 287 BC, it was agreed that all resolutions of the Assembly would be made law. The patrician order, however, continued to hold the upper hand in the Senate. In 14 AD, the Assembly lost its right to choose the magistrates and thus most of its influence.

GOVERNMENT OF THE PROVINCES

Roman territories outside Italy were organized into provinces, each under the rule of a governor who was appointed by the Senate from among retiring praetores and consuls. Some were under the direct rule of the Senate, through a proconsul (Julius Caesar was proconsul of Gaul before making himself dictator).

Governors

The governor was in charge of the courts and also of the armed forces. He was responsible for collecting taxes to be sent to Rome, for putting down any riots and for defending Rome's frontiers. Most governors served in the post for between one and three years.

The governor was assisted by a junior official called a *quaestor* and three or four lieutenants called *legati*, as well as a host of clerks and messengers called *apparitores*. On a more local level, existing officials were often permitted to keep their posts after a territory was conquered. Conquered peoples were generally allowed to keep their religions and customs, as long as they also showed due reverence to the emperor of Rome. The Romans themselves believed that they were benefiting the barbarian peoples they conquered by introducing them to more civilized ways. Their rule, however, could also be extremely ruthless and cruel, and in times of trouble hundreds of executions might take place on a single day.

LIFE UNDER THE ROMANS

Initially, people from the provinces did not enjoy the rank and rights of citizens. Later, however, they were allowed to purchase Roman citizenship and, from the reign of Hadrian onwards, to serve as auxiliaries in Roman armies, assuming responsibility for guarding borders among other supporting roles.

In later times, the emperor often assumed command of troubled provinces, while the financial affairs of the provinces were handled by imperial agents, who were called *procurators Augusti*. It was their job to report any mismanagement by governors back to Rome.

Unpopular leaders

Rome's provincial governors received no salary, but many used their position to make themselves rich and thus often became very unpopular.

LAW

The ancient Romans developed a sophisticated legal system, which in due course was to become a model for many modern systems. The first set of laws, known as the Twelve Tables, was published at the insistence of the plebeian Assembly of Rome in 450 BC. It covered many subjects, from money and property to public order offences. Many more laws were subsequently drawn up by the Senate and, in imperial times, by the emperors. If a case was not directly covered by an existing law, it was up to the judge to interpret how the case should be decided, and his judgement (edictum) then provided a pattern for later cases of a similar type. Ultimately, the emperor Hadrian ordered the collection of the many interpretations of the law made by judges around the empire and had them standardized to reduce the risk of legal decisions varying between regions.

TRIALS

Trials took place before courts presided over by a judge appointed by one of the praetores or (in the provinces) by local governors. People making accusations brought the person they suspected to the court building (often in a basilica near the Forum) and presented their case to a jury of as many as 75 people. Those accused sometimes hired a lawyer (*advocatus*) to represent their defence. The jury decided if the accused person was guilty and the judge pronounced sentence. If a citizen in the provinces felt he had been treated unjustly, he could appeal his case to the judges in Rome.

PUNISHMENTS

Punishments ranged from the payment of fines or compensation for many relatively minor offences to imprisonment, exile or even death. Serious crimes might be punished by flogging or confiscation of property and the loss of citizenship. Another option open to the courts was to sentence someone to work in the mines or as an oarsman in the navy, punishments that in practice often amounted to a death sentence. Judges only imposed the death penalty for very serious crimes. Some condemned prisoners were beheaded, while others were sent into the arena to fight as gladiators or to be thrown to wild animals.

Crucifixion

Among the cruellest punishments of all was that of crucifixion, which involved a person being nailed or bound to a wooden cross and left there to die from previously inflicted wounds or from exhaustion. The victim's legs were sometimes broken to hasten death. Crucifixion was reserved for people who did not hold Roman citizenship, usually slaves or people guilty of the worst crimes.

PART SEVEN

Life in ancient Rome

Through a combination of personal accounts and archaeological remains, we have a rich and detailed knowledge of how the ancient Romans lived, from the lives of soldiers on distant frontiers or plebeians in the city of Rome to those of the emperors themselves. Through conquest, many features of the ancient Roman way of life came to be copied within the empire and their legacy remains strong in many cultures today.

Roman lifestyle

The lifestyle of ordinary Romans remains a source of fascination 2000 years later.

RURAL LIFE

The greater part of the population of the Roman empire lived in the countryside and made their living from farming. Italy itself has fertile soil and a favourable climate for growing crops, and indeed the wealth of many of Rome's great families was originally based on agricultural produce. Even after settling in the city of Rome, many rich people maintained large country estates with fine villas to which they could retreat in the heat of summer.

As well as cereal crops, Roman farmers tended vines, olive groves and livestock. They also grew fruit and vegetables and kept bees for honey. Most of the labour was done by hand, although farmers in Gaul invented a wheeled plough and others built copies of a wheeled threshing-machine (possibly of Carthaginian origin). Men, women, slaves and children all took a share in the backbreaking agricultural work.

Rural idyll

Roman city-dwellers traditionally thought of the rural life as being idyllic, and celebrated its virtues in songs and poems. The dream of many a Roman soldier was to retire to farm a patch of his own land.

REORGANIZATION OF FARMS

Many small farms were ruined during the ravages of the Carthaginian Wars of the third century BC, after which wealthy landowners took the land over and set up much larger, more efficient farms, manned by slave labour. Many impoverished farmers sought a new life in the towns. In 133 BC, the tribune Tiberius Gracchus suggested that public land seized by rich landowners be given to the landless poor, but the Senate objected and both Tiberius and, subsequently, his brother Gaius were killed in riots.

There was a marked decline in agricultural production during the third century as the economy of the empire faltered and collapsed. Large areas of land were left uncultivated and many farms were abandoned. The system of farms based on large villas broke down completely in some parts of the empire and few survived.

SUPERSTITIONS

Country people were more superstitious than urban dwellers and put much faith in local *numens* (guardian spirits). They also made sacrifices to Jupiter, who was believed to control the weather, and other gods, including Mars, the god of farming. Hunters similarly sacrificed to Diana, the goddess of the hunt, before setting out on an expedition.

URBAN LIFE

Urban living was to a large extent a Roman invention. Under Roman occupation, former tribal capitals throughout Europe (often situated on hilltops) were transformed into big new towns, while elsewhere substantial new settlements were founded on major transport routes. Italy itself boasted over 400 Roman towns. Many towns, such as Colchester in England (set up on the site of a former legionary fort in AD 49), were established as colonies for retired soldiers or grew up beside military bases. Early Roman towns grew up in a haphazard fashion, but most later ones had a similar layout, which was based on that of Greek towns, with groups of buildings being constructed on a grid pattern in square blocks (*insulae*) between long, straight streets laid out at right angles to each other.

COMMON FEATURES

Features of Roman towns usually included a forum (the marketplace and social centre of the settlement), a town hall (*curia*), temples, amphitheatres, baths, a gymnasium, a library, warehouses and private housing as well as inns (*tabernae*), eating places (*thermopolia*), bakeries (*pistrina*) and a range of other shops. Bigger towns also boasted monuments such as statues and triumphal arches.

WATER SUPPLY AND DRAINAGE

The Romans understood the importance of a good water supply and an efficient drainage system. They built impressive aqueducts to bring water right into the centre of their towns and cities, and also constructed labyrinths of underground sewers, many of which remain in use today.

TOWN DEFENCES

Roman towns were originally protected by defensive walls, with access through four or more gates. Such defences were considered unnecessary during the early empire, as the presence of formidable armies nearby deterred attack, and the emperors actually prohibited building new walls. Later, however, as invaders threatened, earthworks and stone walls were rebuilt.

Crowded streets

Living space was often cramped within the defensive walls and the roads were very busy. Many Romans complained of the constant noise and of the difficulty in getting about the streets, which were filthy and crowded with carts and pedestrians. At various times the authorities tried to limit the movement of wheeled traffic in busy quarters, but without much success.

Layout of a typical Roman town

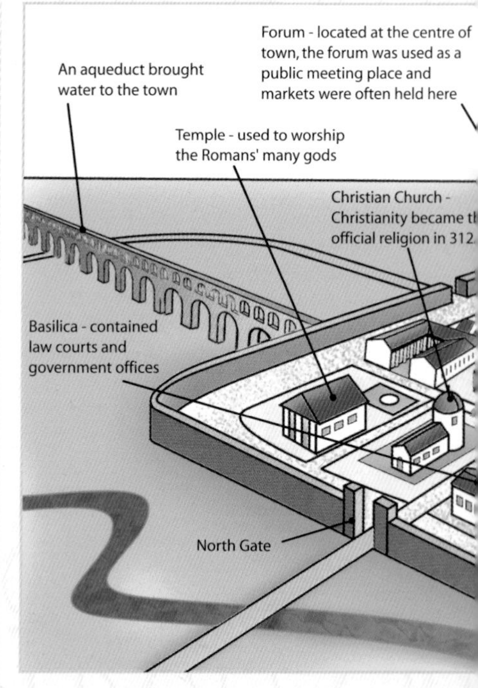

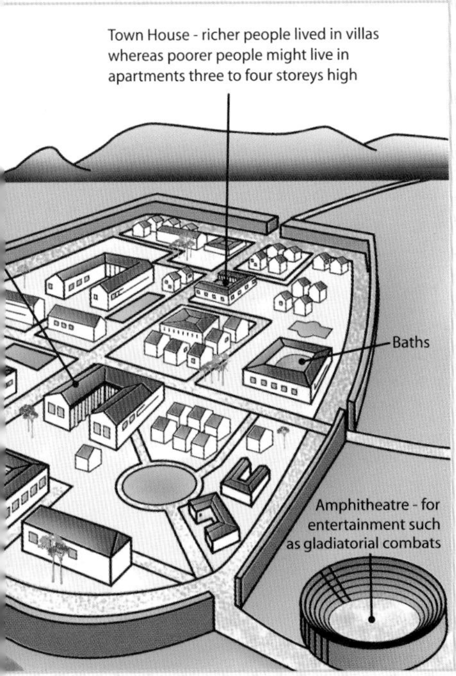

Town House - richer people lived in villas whereas poorer people might live in apartments three to four storeys high

Baths

Amphitheatre - for entertainment such as gladiatorial combats

TRADE

The Romans were a great trading people, and the *Pax Romana* that lasted for most of the first two centuries AD enabled trade to flourish throughout the empire. The suppression of pirates and the building of an extensive road network encouraged merchants to buy and sell goods in distant markets, transporting produce both by road and by sea (which was cheaper).

TYPES OF GOODS

Grain (from Africa, Egypt, Sicily and elsewhere) was the most vital import, essential for feeding Rome's hungry population. Other goods brought back to Rome were wine, olive oil, honey, salt fish, wax, pitch, red dye, black wool and fine cloth from Spain; wine from France; corn, cattle, hunting dogs, pearls, salt and tin from Britain; glass and cloth from Syria; shoes from Greece; incense from Arabia; marble from Africa and Asia; amber from the Baltic; gems from India; and silks from the Far East.

Weights and measures

Roman traders used a variety of scales to weigh produce. Similar balances and steelyards (which were regularly checked by market officials) are still in use today.

CRAFTSMEN AND SHOPKEEPERS

Most craftsmen and shopkeepers throughout the Roman empire came from the lower orders. Potters, carpenters, metalworkers and other craftsmen generally worked in small workshops in their own homes. They sold what they made from a stall that opened onto the street, competing for custom alongside the stalls of butchers, bakers and other shopkeepers.

In many Roman towns, trade was based on the forum, although larger cities such as Rome boasted separate marketplaces where scores of traders sold their wares. Trajan's market in Rome was among the largest of these, housing markets and libraries as well as its own forum and basilica.

The working class

The wealth of Rome was greatly increased by its success as a trading power, but such activity was nonetheless felt to be largely the province of freedmen and beneath the dignity of the patrician class. A senatorial order of 218 BC went so far as to ban patricians from owning vessels capable of carrying more than 300 amphorae. Many traders made fortunes from their businesses. Merchants often worked in co-operation with each other as members of large trading companies or guilds, which protected their collective interest.

MONEY

The Romans learned the use of coins from the Greeks, and established their first mint around 290 BC. Before then, commerce was based chiefly upon the exchange of goods of roughly equal value. The introduction of money made trade much simpler. It was convenient for the payment of taxes and the salaries of Rome's soldiers as well as for ordinary everyday transactions.

COINAGE

The first coin minted in the republican era was the bronze ass. Later came the silver didrachm and the silver denarius (worth between ten and sixteen asses), the silver sestertius (worth between two-and-a-half and four asses) and the aureus, Rome's first gold coin.

Various changes were made to the coinage under the empire in the face of repeated price rises and the increasing cost of precious metals. The coin of the least value now became the quadrans, a copper coin worth a quarter of an ass (now also made of copper). The bronze semi was equivalent to two quadrantes or half an ass, while the dupondius was worth half a sestertius. The sestertius (worth four asses) was now made of bronze and worth quarter of a silver denarius, while the gold aureus was equivalent to twenty-five denarii.

A single quadrans was the fee charged to enter the public baths, while a craftsman might expect to earn around one silver denarius a day.

Portraits of emperors on Roman coins promoted acceptance of the coinage throughout the empire, although ultimately a combination of rising prices and increasing numbers of forgeries persuaded many Romans to go back to a system of bartering.

LOANS AND TAXES

The Romans developed a sophisticated financial framework to control trade and economic activity. Some enterprising citizens became hugely rich through the banks they founded, providing loans to people who needed money.

The state acquired vast sums through taxation, especially from the provinces. Anyone who bought or sold slaves, property or other goods had to pay tax on their purchases to official tax collectors. The idea of paying taxes did not appeal to ordinary Romans. The first taxes were imposed to meet the needs of the state in time of emergency, but gradually multiplied in number. The unpopular property tax was abolished in 167 BC as large amounts of plunder from Greece reached Rome, and as more territories came under Roman occupation the burden of paying tax shifted to Rome's foreign possessions.

TRAVEL AND TRANSPORT

Good communications and reliable means of transport were essential to Roman prosperity and security. Roman armies needed to be able to move quickly from one part of the empire to another, while merchants and ordinary citizens also needed easy access to remote markets and communities.

ROMAN ROADS

The Romans rank among the world's greatest ever roadbuilders. Their impressive network of roads was constructed by the army, specifically to speed up troop movements. The first Roman road was the Via Appia, connecting Rome and Capua. It was begun in 312 BC and completed over 100 years later. Later roads connected Rome with the most remote parts of the empire, inspiring the saying 'All roads lead to Rome'.

Roads of the giants

The roads built by the Romans remained the best available well into medieval times and became the basis of many modern transport routes. Later generations, who had long lost the skills necessary to construct such roadways, believed the roads of the Romans to have been the work of giants.

By the second century AD, the network comprised over 80,000 km (50,000 miles) of roadway.

Construction

Roman surveyors used basic surveying devices to plot straight courses from one landmark to the next. Legionaries first cleared the proposed route of turf and trees and then dug a trench in which they laid a base of logs and stones. On top of this were placed further layers of stones, concrete and broken tiles before a final hard-wearing surface layer of stone slabs or cobbles was added. To prevent water damage, roads were constructed with a camber (hump), so that rainwater ran off into ditches beside the roadway. Valleys and rivers were spanned by bridges and viaducts (many of which still stand), and milestones were erected every mile (equivalent to 1000 paces). Building such roads was expensive, costing over 100,000 sesterces per mile.

Journeys by road

Travellers walked, rode on horseback or on donkeys, or were carried in carts or wagons. A person travelling by cart on a Roman road could expect to travel about eight miles in a single day. By contrast, the emperor Tiberius, travelling non-stop using a relay of chariots, once covered 500 miles in just twenty-four hours. Long journeys were made easier by the building of *tabernae* (inns) and mansions (guest houses) at regular intervals along the way.

Road network

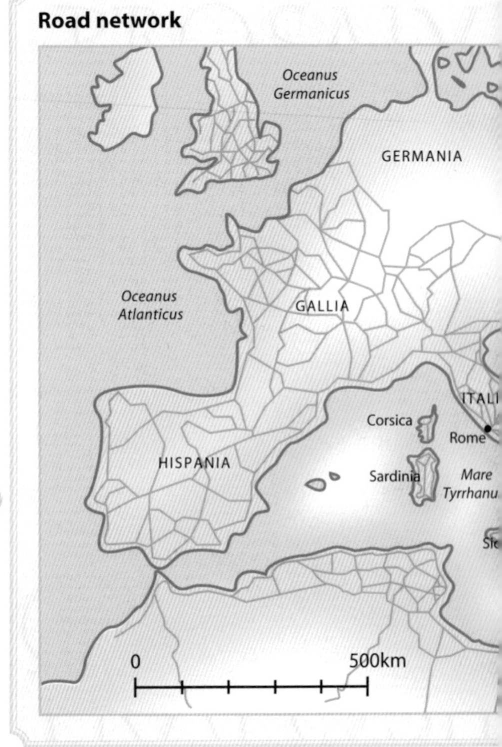

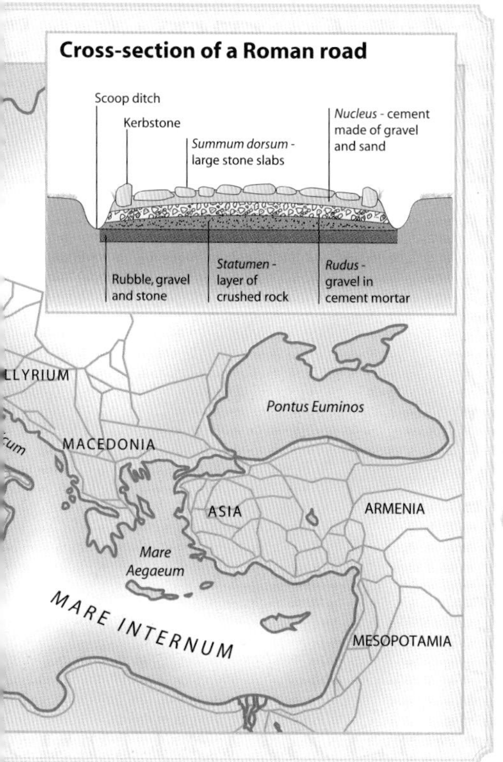
Cross-section of a Roman road
Scoop ditch
Kerbstone
Summum dorsum - large stone slabs
Nucleus - cement made of gravel and sand
Rubble, gravel and stone
Statumen - layer of crushed rock
Rudus - gravel in cement mortar
LLYRIUM
MACEDONIA
Pontus Euminos
ASIA
ARMENIA
Mare Aegaeum
MARE INTERNUM
MESOPOTAMIA

TRAVEL BY SEA

The Romans were not historically a seafaring people, but after the third century BC a substantial navy was built to fight Carthage. Having achieved mastery of the Mediterranean, they built fleets of merchant ships to transport cargoes great distances to and from Rome.

Rome was served by the nearby port of Ostia at the mouth of the Tiber. Incoming ships brought wine and grain as well as tribute and taxes from the provinces, wild animals for the city's circuses and other cargoes. Goods were unloaded and transported on barges up the Tiber to Rome. Soldiers, traders and citizens were also transported by sea to and from the provinces, thus promoting the spread of Roman culture. Lack of navigational knowledge initially obliged ships to stay close to coasts, but as the Romans became more skilled they ventured further out to sea, finding their way by observing the position of the sun and the stars. They also built lighthouses to guide ships safely into port.

Amphorae

Wine and olive oil were transported in large pottery jars called amphorae. Amphorae used to carry olive oil were smashed after use as the oil soaked into the pots and went rancid. The broken pieces were then used for ballast.

Shipping

From the evidence of wrecks, it appears that Roman merchant ships were solidly built, with wide hulls that allowed for large cargo holds. They came in a variety of sizes, and were propelled by a large square mainsail and one or more smaller sails. Some Roman ships had elaborately carved prows and sterns.

This rare carving showing a Roman ship comes from the funeral stone of a Roman ship builder.

FAMILY LIFE

The family lay at the core of ancient Roman life. Each family was headed by the paterfamilias, the senior male, who presided over family rituals and had control over all other family members (wives, children, slaves and property of their sons). When he died, his sons became the heads of their own families, thus forming a clan (*gens*) of related families. In practice, however, the day-to-day running of households was left to the wives.

DAILY ROUTINE

Most Roman households rose before dawn. After a light breakfast, the man of the house put on his toga and presided over prayers at the household shrine, asking for protection both for his family and for the emperor. Any babies were fed by a wet-nurse, while other children were expected to wait quietly for the arrival of their Greek tutor.

Household pets

Many Roman families kept animals. Some were working animals, such as hunting dogs, guard dogs or horses. Others were simply pets, some of which wore collars bearing the owner's name, in case they went astray.

Boys left for school while it was still dark, escorted by a family slave, often buying breakfast on the way. At this early hour, rich men, called patronae, received their clientes – individuals who relied upon their patron for small gifts of food or money (perhaps six sesterces a day) in exchange for their political and social support. They then left for the law courts or the Senate.

Ordinary citizens went into the crowded streets to visit the public latrine, chat with their neighbours or see their patron. Some had posts in the financial or administrative sectors, or pursued a career in politics or the army, but few citizens undertook physical work, which was left to freedmen and slaves. Poorer citizens earned a living as craftsmen or shopkeepers. After a modest lunch, many Romans visited the public baths, where they could meet friends, chat, exercise, read or generally relax. The main meal of the day was eaten in the evening. Women spent most of their time busy with household tasks.

THE ROLE OF WOMEN

Women in ancient Rome had little control over their own lives, and what became of them depended a great deal on the wealth and status of their fathers, husbands and brothers. Females were commonly believed to be the weaker sex, and the birth of a girl was often a cause for lamenting (some fathers hated their daughters or left them as babies outside to die of cold).

The chief role of women was to marry and bear children. Given the limited medical understanding of ancient Roman doctors, childbirth was a hazardous activity, and many women died giving birth.

Unless chosen to serve as priestesses, women took little part in public life beyond supporting male relatives, rarely leaving the confines of the family home. Girls from even modestly well-off families were not expected to have careers, although the situation improved after the reign of Augustus and a select few were allowed to work as teachers, doctors, midwives or businesswomen. As most received little education, if any at all, this was no easy achievement. Most women were expected to devote themselves to managing the household (though rich women usually had slaves to do any actual work).

Poor women

Women from less wealthy families often had more freedom than women from richer homes, although

Imperial wives

The wives of emperors and other wealthy and powerful men had more opportunity to influence affairs of the day. Among the most famous (or notorious) of imperial wives was Agrippina, who murdered her husband Claudius in order to make her son Nero emperor.

their lives tended to be much less comfortable. Despite the fact that they were unlikely to receive much formal education, they were often expected to support their families by taking paid work. Many women worked in markets or on stalls, as bath attendants, as weavers, shepherdesses or needlewomen, among other roles. There were also thousands of female slaves, whose duties ranged from household maids to farm labourers.

LOVE AND MARRIAGE

Most marriages were arranged for financial, political or social advancement rather than love, although this was not always the case. Parents might choose husbands for their daughters when they were as young as twelve, though most did not marry until fourteen. One reason for such early marriages was that younger women were thought more likely to survive the perils of childbirth. Under the republic, everything a girl owned became the property of her father-in-law when she married. Later, however, the law was changed to allow a woman to retain some control of her own property.

Wedding ceremonies

The date for a wedding was chosen with care to avoid the many unlucky days in the Roman calendar. The latter part of June was considered the best time of all. The ceremony itself took place at the bride's family home, after appropriate sacrifices had been made and the priests were satisfied that the omens were favourable. The bride wore a white wedding dress, a headdress of flowers, an orange veil and red shoes. Vows were exchanged and a marriage contract signed. The bride received a ring and the hands of the happy couple were joined by the bridesmaid. The wedding party then enjoyed a banquet before accompanying the newlyweds in procession to the groom's house with music and burning torches. When she arrived at her new home, the bride carried out certain rituals and might be carried over the threshold by her husband. The cost of weddings became so high that eventually the emperor Augustus passed a law limiting spending on weddings to 1000 sesterces. The law was widely ignored.

CHILDREN

Limited medical knowledge meant that the ancient Romans had to endure a high infant mortality rate, with relatively few children surviving to adulthood. When a baby was nine days old, a naming ceremony took place and the infant was given a charm known as a *bulla*,

Divorce

Unhappy marriages often ended in divorce, particularly among the nobility. Several emperors, such as Tiberius, divorced their wives for political reasons.

Weddings were important social events, often celebrated with sacrifices (here a bull).

which was believed to keep it safe from evil spirits until it grew up. Children were expected to behave like adults and were usually dressed in tunics not unlike those of their parents. If born into poor families they were also expected to work alongside their parents from an early age. Those children who were born into wealthier families had an easier time, being tended by their family slaves and, on occasion, accompanying their parents to official ceremonies. Such children were sent to school from the age of seven, and might also be taught at home by tutors. Few girls received a formal education, being taught instead domestic skills that would be of use to them when they had their own households to run.

Toys

Roman children were given a range of toys to play with. These included seesaws, whips and tops, kites, rag dolls, toy houses, model animals (often made of poisonous lead) and marbles (made of glass and pottery). They were expected to offer their toys to the household gods on getting married.

Adulthood

Most children were officially declared to be adults around the age of seventeen. When the time came, they were taken to the forum for a formal ceremony, during which they discarded their childhood clothes and *bulla*. Boys were then presented with the toga of an adult, were given their first shave and registered as citizens. There was then a party to celebrate.

Roughly a third of all Roman children died in infancy, often of malaria or other diseases.

EDUCATION

Children born into poorer families received little or no education and never learned to read and write, but boys from wealthier backgrounds could expect to be sent to a *ludus* (an elementary school) from the age of six or seven and to remain there until they were eleven years old. Girls of a similar age might receive some basic education at home, although this tended to be limited to learning domestic skills.

THE SCHOOL DAY

The school day started at dawn and ended at midday. Pupils were escorted to school by a family slave, known as a *paedagogus*, who remained with them throughout their lessons to ensure that they behaved correctly.

Writing tools

Written exercises were done on wooden tablets coated with wax, using a stylus (a metal pen), or else by scratching letters on broken bits of pottery. Older pupils might use pens of reed or metal to write on sheets of papyrus (a form of paper made from Egyptian papyrus reeds and rolled into scrolls). The ink they used was made from soot mixed with water and other ingredients.

Most schools were situated in sheltered public areas, in private houses or in rooms behind shops, with the children being taught in classes of around a dozen pupils. The teachers were often Greek slaves. Among the most famous teachers was Magister Perry, who taught four future emperors and is credited with the invention of homework.

As well as learning how to read and write, pupils were given instruction in mathematics and introduced to the writings of famous authors. The children were required to learn much of what they were taught by heart, and those who did not risked a thrashing.

FURTHER EDUCATION

For most Roman boys, formal schooling ended at the age of fourteen, but for many others their education continued at home, where they were taught by a private tutor called a *grammaticus*. Their studies now expanded to include history, philosophy, geography, music, astronomy and Greek. These studies prepared them for the final stage of their education under a *rhetor*, a teacher who taught rhetoric (public speaking).

Such knowledge was thought essential for anyone who planned a career as a politician or lawyer. Training in the arts of rhetoric might last for many years (the great speaker Cicero continued his until he was thirty).

HOUSES

Most inhabitants of Rome during the later republican and imperial eras lived in apartment blocks which were called *insulae* (islands). A survey of 350 AD revealed that there were 46,602 *insulae* in Rome, compared to 1790 private houses. Insulae could be up to six storeys high, although because of the risk of collapse most were restricted to just four or five storeys.

INSULAE

It was expensive to rent a *cenaculum* (an apartment) within an *insula* (up to 30,000 sesterces a year) and most people were crammed into small rooms with few luxuries. The best apartments, on the lower floors, were richly decorated and well-furnished. Poorer people lived on the cheaper upper storeys, while the ground floor spaces were usually rented out as shops or inns. Most *insulae* were badly built and poorly maintained.

Furnishings

Roman homes were sparsely furnished. The wealthier families owned a range of beds, couches, tables, stools and cupboards. Rooms were illuminated by oil lamps made of pottery or bronze.

Usually the only heating was provided by woodburning braziers, which filled the rooms with smoke and could easily set buildings ablaze. There was no running water and no drains or toilets. The windows lacked glass, and when the shutters were closed the air inside soon became stale.

HOMES OF THE WEALTHY

Only the rich could afford their own *domus* (private house). Most of these shared the same basic design, with an atrium (hallway), in which guests were received, opening into an airy central courtyard. Other rooms included sleeping quarters, a *tablinum* (study) and a *triclinium* (dining room). Among other features might be a *lararium* (family shrine) and a *peristylium* (small garden). In the first century AD, the Romans introduced the hypocaust, a system of central heating by means of which rooms were warmed from below by warm air circulating in the basement.

Country villas

The wealthiest Romans owned both a townhouse and a country villa, to which they retreated during the summer months when the heat and the threat of disease made the cities unpleasant and dangerous places. These country villas were often large, with many elegant rooms which were decorated with frescoes, mosaics and marble.

Layout of a typical Roman house

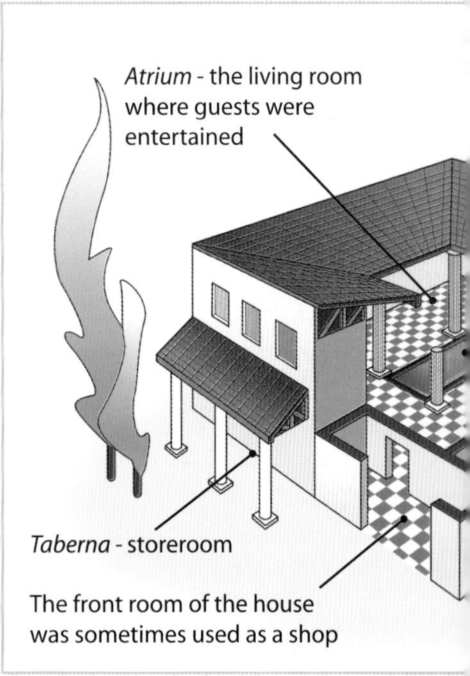

Atrium - the living room where guests were entertained

Taberna - storeroom

The front room of the house was sometimes used as a shop

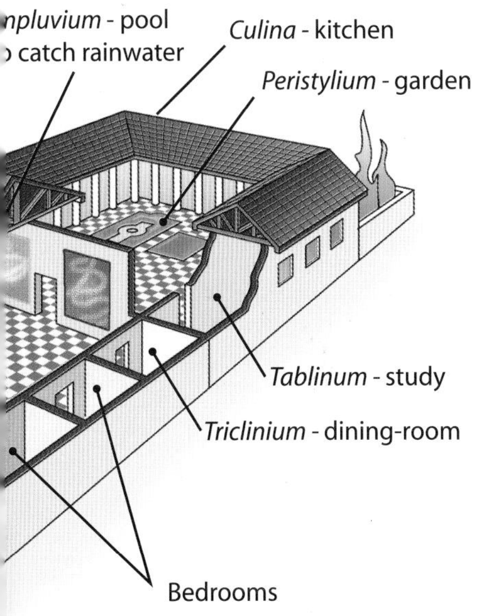

npluvium - pool
catch rainwater

Culina - kitchen

Peristylium - garden

Tablinum - study

Triclinium - dining-room

Bedrooms

ROMAN BATHS

Personal hygiene was important to the ancient Romans, but as most private homes lacked bathing facilities, citizens regularly visited the public baths that were a feature of virtually all Roman settlements, large and small. The first major bath complexes (initially privately owned, but later also state-owned) were built during the second century BC. Their number increased steadily; in Rome itself they rose from 170 during the reign of Augustus to more than 900 by 300 AD.

Bath complexes were among the largest and most luxurious buildings in Roman towns, with sophisticated underfloor heating and plumbing systems. As well as bathing facilities, they might include changing rooms, gymnasiums or exercise yards, reading rooms and gardens with outdoor pools. The biggest bath complexes in Rome included the Baths of Diocletian which, when completed in 305 AD, could accommodate 3000 bathers, and the Baths of Caracalla.

Getting clean

The Romans did not use soap, but instead washed themselves with olive oil, which they then scraped off, along with any dirt, using special tools called *strigils*.

PUBLIC BATHING

Most Roman citizens went to the baths in the afternoon, and might stay for several hours. Women had separate baths or went in the morning. After undressing, bathers visited the *tepidarium* (a warm room with a small pool), followed by the *caldarium* (a hot room with another pool, where the heat made bathers sweat profusely, thus clearing the pores of the skin) and, finally, the *frigidarium* (which housed a refreshingly cold pool). Invalids were encouraged to visit an even hotter room, called the *laconicum*. After bathing, visitors might enjoy a massage, a haircut or some form of beauty treatment. They could also buy drinks and snacks and spend time exercising, playing board games or socializing with their friends.

Some Roman baths, such as those at Bath, England, took advantage of natural hot springs.

FOOD AND DRINK

The ancient Romans enjoyed a mixed diet, with a wide variety of fruit, vegetables, poultry, pork, fish and other foodstuffs available at market stalls. However, they lacked some of the ingredients commonly eaten in Italy today, such as potatoes, tomatoes, rice and pasta.

Most people ate little during the day. Breakfast might be no more than a drink of water, though some began the day with a light snack of bread or wheat biscuits with honey, dates or olives. Lunch (*prandium*) consisted of similar fare. The main meal of the day was dinner (*cena*), which was eaten in the afternoon or evening. For poorer people this was often no more than wheat porridge. As only the largest houses had their own kitchens, slightly better-off urban citizens relied on hot food sold from bars or other eating-places, dining on such dishes as sausage in semolina, bacon and beans or bread, beans and lentils with a little meat.

DINNER PARTIES

Richer citizens enjoyed a wider choice of dishes, and frequently held or attended dinner parties. Under the republic, only men were invited to these, but later women were also included among the guests. Dinner parties might start in the early afternoon and last until past midnight. The diners reclined on couches grouped in threes around a low dining table, the fourth side being left clear for slaves to serve the courses. Diners did not use knives and forks, but instead used their hands (or, occasionally, spoons) to eat.

Courses and wine

There were usually three main courses. The first was an appetizer, which might consist of raw vegetables, eggs, fish, olives, salad, mushrooms or shellfish, washed down with *mulsum* (wine and honey). The main course might comprise as many as seven dishes, including fish, meat or poultry, all prepared with vegetables and sauces. Exotic recipes featured roasted songbirds, stuffed dormice cooked in honey and poppyseeds, and lamb's innards stuffed with sausagemeat. Dessert might be fresh or dried fruit seasoned with pepper. Diners often drank a great deal of wine, and such evenings could easily end in drunken revelry or even in an uninhibited orgy. More often, the entertainment offered during the meal ranged from intellectual conversation to music, dancing, acrobatics and poetry readings.

Overeating

The habit some ancient Romans had of deliberately making themselves sick at dinner parties to make room for more rich food was probably confined to only the greediest diners.

CLOTHING AND HAIRSTYLES

The citizens of ancient Rome mostly wore clothing that was made of wool or linen. These changed in style surprisingly little over the centuries.

Men of all social classes typically wore a loincloth under a simple sleeveless tunic, which comprised two rectangles of cloth stitched at the sides and shoulders and secured with a belt at the waist. Over their tunic, wealthier citizens wore a longer garment called a toga, which comprised a large piece of (probably) semicircular cloth that was wrapped around the body and over the left arm. The toga was usually white, although there was some use of dye. The emperor was the only person allowed to wear a purple toga, purple dye (obtained from murex seashells) being the most expensive dye available. Senators, meanwhile, wore togas with a purple stripe. Trousers were considered unmanly, but were commonly worn in cooler regions of the empire, as were cloaks.

Footwear

Both sexes wore leather sandals of various designs, some with strapping up the lower calf. Men also wore heavy boots for outside work, with soles studded with nails.

WOMEN AND CHILDREN

Women wore a loincloth and, sometimes, a bra under a long tunic of fine wool, linen or (rarely) Chinese silk. Over this they wore a *stola* (a long robe) and *palla* (cloak). When outside, they might also wear a veil or scarf over their hair, or pull their palla over their head.

Roman children wore tunics like their parents. However, at around the age of fourteen, boys exchanged their childhood toga (the *toga praetexta*) for that of an adult (the *toga virilise*).

HAIR AND BEARDS

Hairstyles changed over the years. Men (who had their hair cut at a barbershop) generally wore their hair short, sometimes oiled and curled at the temples. Beards came into fashion at various times, being widely worn, for instance, in the early republic and again (in short and clipped form) around the middle of the second century AD.

During the period of the early republic, women often wore their hair in simple buns. Later, however, they favoured more elaborate hairstyles and had their hair braided and curled by their slaves. By the fourth century AD, Roman women had reverted to a simpler style, with their hair tied back and a centre parting.

MAKE-UP AND JEWELLERY

The ancient Romans took care over their appearance. As well as wearing elegant clothes and elaborate hairstyles, fashionable citizens also used make-up and adorned themselves with items of jewellery. Wealthier women could call upon the services of a professional hairdresser, called an *ornatrix*, who was also skilled in the application of make-up.

MAKE-UP

It was considered fashionable for wealthy women to wear pale make-up, whitening their faces with powdered chalk or even white lead (a dangerously poisonous substance). For making blusher, they used the sediment of red wine or a plant dye called *ficus*.

Other preparations included eye make-up (based on ash or antimony) and facepacks (consisting of bread and cream). Perfumes and other cosmetic preparations, which were often prepared by family slaves, were stored in an array of pots and bottles.

Mirrors

To check their appearance before being seen by others, Roman women used mirrors of highly polished metal, as reflective glass mirrors had yet to be invented.

JEWELLERY

Ancient Romans of both sexes wore *fibulae* (brooches or pins) to fasten their tunics and togas. Both men and women also wore rings on their fingers. Originally only high-ranking aristocrats were allowed to wear gold rings, but, later on, as gold mined in the provinces became increasingly available, such rings were worn more widely. Rings set with specially carved stones were often used to seal documents, while others were worn to ward off bad luck.

Other items of jewellery included gold and silver necklaces, bracelets, earrings, anklets and hairpins. Also popular were cameos comprising miniature carvings in semi-precious stone (usually sardonyx), which were commonly worn as brooches or medallions. Many items of jewellery incorporated precious gems, such as opals, emeralds, sapphires and pearls, as well as ivory. Poorer people wore jewellery decorated with glass or bronze.

Beauty treatments

Among more outlandish beauty treatments was a preparation of rats' dung, pepper and other ingredients that was rubbed into the scalp to prevent baldness. Some people wore false teeth specially imported from Germany.

LEISURE ACTIVITIES

Because most manual work was done by slaves, Roman citizens had plenty of opportunity for indulging in leisure activities. As well as trips to the baths, attending dinner parties, hunting, fishing, wrestling, exercising and playing dice and other games, they could choose from a wide range of free public entertainments.

BREAD AND CIRCUSES

The emperors knew that keeping Rome's population content with a plentiful supply of food and entertainment ('bread and circuses') would help maintain public order. As well as providing a monthly food handout (*annona*), they arranged theatrical performances (*ludi scaenici*), chariot races (*ludi circenses*), gladiatorial contests and wild beast shows (*munera*). Many entertainments were bloodthirsty, involving the death of hundreds of wild

Arenas

Vast stadiums and circular amphitheatres with many rows of tiered seats were built throughout the empire to accommodate massive crowds of spectators at such events. Some, like the Circus Maximus in Rome, could seat more people than the largest of modern football stadiums.

animals, slaves, convicted criminals and gladiators. Some events were staged on a huge scale and lasted many days. The opening of Rome's Colosseum was marked by 100 days of games in which 9000 animals were killed. On other occasions arenas were flooded with water so full-scale naval battles could be staged.

A typical day at the games might begin with *bestiaries* (huntsmen) in combat with wild animals, such as lions, leopards, bears, tigers, rhinoceroses and elephants. Crowds might revel in the gory sight of condemned criminals or Christians being thrown into the arena to be torn apart by wild beasts. At midday any bodies were removed before the arrival of the gladiators.

Half-starved wild animals, which savaged to death scores of victims, were a major attraction in arenas.

THE GLADIATORS

The stars of the Roman games were the gladiators, who met in single combat and fought to the death in front of vast crowds. Although some were paid volunteers, most were slaves or criminals who accepted life in the arena as an alternative to execution. They were trained in special gladiator schools and, if successful enough, might eventually be rewarded with their freedom. Most gladiators, however, could only hope for a brief life of celebrity before a violent and bloody death.

Types of gladiator

There were four types of gladiator, each armed with different weapons and armour. A Thracian (originally a prisoner-of-war from Thrace) was a swordsman who fought with a curved sword and small shield. A *murmillo* – from the Greek *mormylos* (fish), a reference to the fish-shaped crest on his bronze helmet – carried a sword

Thumbs up

The traditional notion that emperors communicated their verdict on a defeated gladiator by raising or lowering their thumbs is inaccurate. In reality, it seems that they covered their thumb with the other four fingers to indicate they wished to spare the defeated man or pointed their thumb sideways (like a sword) if they wished the man to die.

and large shield and wore armour on his sword arm. A Samnite (named after the warlike Samnites with whom the Romans fought wars in the fourth century BC) also carried a sword and shield. A *retiarius* (net man) carried a weighted net, in which he tried to tangle his opponent before killing him with his three-pronged trident.

Death in the arena

The gladiators' entrance caused great excitement. Music played, and before the contest began the gladiators acknowledged the emperor with the traditional salute 'We who are about to die salute you'. They fought either in pairs or in groups. When wounded, a gladiator could appeal for mercy. If the crowd approved because he had fought well, the official in charge might spare him. If mercy was denied, the loser was killed sometimes by an official dressed as Charon, the legendary ferryman to the Underworld, who clubbed the warrior to death.

THE COLOSSEUM

Some of the great gladiatorial amphitheatres still stand today. Most famous of all was, and still is, the Colosseum in Rome, which was opened in 80 AD by the Emperor Titus and held 50,000 spectators. In hot weather, they were sheltered from the sun by a huge canvas awning.

Overleaf: A *murmillo* gladiator appeals to the crowd for their verdict on a defeated opponent.

CHARIOT RACES

Most popular of all the public events that were staged in ancient Rome were the chariot races. Vast crowds packed *hippodromes* (racetracks) throughout the empire to watch up to twenty-four races in a single day. Most charioteers (who were usually slaves) drove either a *biga* (a two-horse chariot) or a *quadriga* (a four-horse chariot), although six or even as many as eight horses were sometimes used. The chariots themselves, which were deliberately light and designed for speed, were sometimes elaborately decorated.

Up to twelve chariots raced anticlockwise around a broadly oval track, which was split down the middle by a barrier of pillars called the *spina*. The most dangerous parts of the track were the two bends around the ends of the *spina*, where chariots often overturned. Injuries and even deaths among the charioteers were not infrequent. The winner of the

The Circus Maximus

The oldest and largest of all the racetracks was the Circus Maximus in Rome. This vast stadium, which had a racetrack 550 metres (1800 feet) long, was capable of holding up to 250,000 spectators. You can still visit it today.

race after seven laps, equivalent to eight kilometres (five miles), received the victor's palm and a purse of gold. The most successful charioteers became rich and famous and might even be rewarded with their freedom from slavery.

Legend had it that it was Romulus, the legendary founder of Rome, who constructed the city's first circus for chariot-racing. By the third century AD, Rome had no less than eight racetracks, and there were many more throughout the empire.

SUPPORTERS

The spectators bet on the four teams that took part in each race and dressed in their favourite team's colour (red, green, blue or white) to show their allegiance; the white team was owned by the emperor himself. In contrast to gladiatorial shows and theatrical performances, men and women sat together to watch chariot-racing. The atmosphere was very lively, with much cheering and intense rivalry between different groups of supporters, sometimes even culminating in outbreaks of violence (for instance, thousands of people died in Constantinople in 532 AD during riots between supporters of the rival blue and green teams). The excitement of chariot-racing was memorably recreated for modern audiences in the 1907, 1925 and 1959 films of *Ben Hur*.

DEATH IN ANCIENT ROME

The ancient Romans lived relatively short lives, which often ended prematurely due to poor diet, disease, lack of medical understanding and generally hard lives. Women frequently died in childbirth, while one in three children died in infancy. One in two people were dead before reaching the age of fifty.

Like the Greeks, the Romans believed that the spirits of the dead were rowed across the river Styx to the Underworld. There they were judged and either sent to heaven (Elysium) or hell (Tartarus).

FUNERAL RITUALS

When a person died, their body was washed and coated with oil. In the case of important people, the body was then laid in state in the atrium of their home, where they

Roadside graves

By law, because of the risk of disease, cemeteries were located outside the city walls, usually lining main roads. The preferred graves were those closest to the road, where passers-by were more likely to notice them and thus keep the memory of the deceased person alive.

could be viewed by friends coming to pay their last respects. A coin was carefully placed under the tongue of the corpse to pay the ferryman to carry their spirit across the River Styx. The body might then be carried on a litter to the forum, accompanied by musicians and, sometimes, *praeficae* (professional mourners). In republican times, the relatives of the deceased wore death masks and mourning robes. A eulogy praising the achievements of the dead person and his or her ancestors was often read out during the funeral procession.

Burying the dead

Following the funeral service, the body was taken either to a cemetery or place of cremation (after which the ashes were placed in an urn and laid in a tomb or underground chamber called a columbarium). The Romans originally burned their dead, but later preferred to bury corpses, often in substantial family tombs. Graves were often located by stone grave-markers bearing the name of the deceased, and in many cases carvings relevant to the dead person's life.

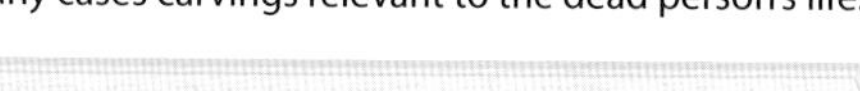

Imperial tombs

The ashes of many deceased members of the imperial family were interred in the impressive Mausoleum of Augustus in Rome. Completed in 28 BC, this massive structure still exists as an imposing ruin.

PART EIGHT

Roman culture

As well as being great soldiers and intrepid traders, the Romans made many significant advances in cultural fields. They built structures that have remained standing for 2000 years, left behind fine mosaics and sculptures, contributed some classic pieces of writing to world literature and expanded upon the achievements of the ancient Greeks in the fields of engineering, philosophy, medicine and the sciences.

The Roman legacy

The impact of Roman culture upon world civilization, from architecture and law to theatre and literature, cannot be overestimated.

CULTURE AND LEARNING

In many respects, Roman culture was based on that of the ancient Greeks. The conquest of Greece in the second century BC promoted close contact between the two civilizations and enabled the absorption of Greek ideas. Many Greek slaves were brought to Rome to pass on their knowledge as teachers, doctors, artists and scientists. The Romans, in their turn, communicated their culture to remote parts of the known world. Conquered peoples were encouraged to adopt Roman ideas and a Roman way of life, so cultural differences between the inhabitants of the empire were reduced and a distinct Roman culture was shared in common throughout Roman-occupied territory.

ARTISTS AND CRAFTSMEN

The Greek influence was particularly strong in arts and crafts. Many celebrated Roman sculptures, for instance, were copies of now lost Greek works. The emperors imported numerous fine works from Greece and Greek colonies in Italy, providing Roman artists with models. Virtually no paintings by Roman artists have survived, but surviving frescoes and fragments of decorated pots give an idea of artistic styles and subjects. The work of Roman craftsmen is also preserved in jewellery and metalwork and in details of Roman architecture.

ARCHITECTURE

The ancient Romans were great architects. As well as temples, villas and other impressive public buildings, they constructed huge amphitheatres, basilicas, arenas and stadiums. In the process, they greatly advanced understanding of civil engineering and materials. The Romans modelled their buildings closely upon those designed by Greek architects, and they often employed Greek architects and craftsmen, who were admired for their mastery of construction. The Romans added their own ingenuity, however, with their innovative use of arches. Their invention of concrete, in particular, enabled them to build some of the largest structures in the classical world, including the first domes.

ARCHES AND VAULTS

The Romans were among the first builders to perfect the use of arches. Arches are capable of holding great weights, making possible buildings, aqueducts and bridges of greater height and scale. The Romans made arches by constructing a wooden support between two stone columns and assembling wedge-shaped stones on the support before finally removing the wooden frame.

Vaults were constructed by lining a series of arches side by side. By making a circle of arches crossing over each other, Roman architects created massive domes, of which the Pantheon in Rome, built in 25 BC, was the largest, remaining unsurpassed in size for many centuries. Such domes also relied upon the use of concrete, a Roman invention dating from the second century BC.

COLUMNS AND CAPITALS

Under the influence of Greek architecture, Roman architects made extensive use of columns in their public buildings. Each column was topped with a separate section called a capital, upon which the lintel rested. There were five different styles of columns. From the Greeks came the Doric style (with a plain, undecorated capital), the Ionic style (with a scroll-topped decoration called a volute) and the Corinthian style (in which capitals were elaborately decorated with acanthus leaves and other designs). To these the Romans added the Composite style (a combination of existing styles) and the Tuscan style (with a plain capital).

Enduring structures

The fact that Roman builders knew their trade is evidenced by the large number of major buildings that remain standing today. Not all buildings were of such quality, however, and many residential houses in Rome and other urban centres were relatively flimsy and subject to collapse.

BUILDING METHODS

The Romans devised a range of technical instruments to help them during building projects. Architects drew plans and made scale models using bronze dividers. Masons and carpenters made measurements with bronze foot-rulers and checked that walls were vertically straight with a plumb bob (a bronze weight suspended on a cord). The builders used a range of chisels and other tools when working on wood and stone.

Scaffolding and cranes

Wood scaffolding was often constructed to facilitate the building of tall structures. The Romans even used cranes, which were driven by slaves in a treadmill, to manoeuvre blocks of stone into position.

Building materials

Attention was paid to the quality of building materials, with much marble being used in important buildings. Double walls were filled commonly with stones and cement, then finished with a coat of plaster. Roman bricks were fired in kilns and were typically small and flat. They were often laid in decorative patterns and cemented in place with mortar made from volcanic ash. Roman walls can be identified by alternating layers of bricks and stone. Tiles and gutters were made of baked clay and were usually stamped with the name of the factory from which they came.

CIVIL ENGINEERING

The Romans were brilliant engineers, building roads, aqueducts, bridges, bath-houses and underground sewers. The road network made it possible to travel to distant parts of the empire in a relatively short time. To span rivers and valleys, Roman engineers became expert bridge-builders. The usual procedure when crossing a river was to construct a temporary wooden bridge resting on a series of moored boats. Stakes were then driven in circles into the river bed and the water thus enclosed pumped out and the space filled with stone blocks upon which a permanent stone bridge could then be built. The introduction of the arch meant that bridges and aqueducts could span much longer distances than was previously possible.

AQUEDUCTS

The Romans realized the importance to public health of a proper water supply. Accordingly, they advanced the science of water systems and sewers. In the first century AD, Rome's needs were met by nine aqueducts carrying 1000 million litres (222 million gallons) of water a day in lead pipes from lakes and streams in nearby hills. Some were on a vast scale: the Aqua Claudia was over 64 kilometres (40 miles) long. The aqueducts fed outlets in barracks, temples, fountains and cisterns from which water was fetched by residents. Some wealthy

The Pont du Gard in southern France remains one of the most impressive feats of Roman civil engineering.

Romans had water piped directly into their homes. Water was also harnessed to drive mills.

Roman cities were served by underground sewers. In Rome, filthy water was carried by streams through a network of tunnels into a main sewer, the Cloaca Maxima, and thus into the Tiber. Citizens were supplied with public latrines, some of which had as many as sixty seats. It appears from what remains of these that the Romans had little concern about privacy, and treated the latrines as a place to sit and chat with neighbours.

SCULPTURE

The ancient Romans placed countless statues in their temples, forums and public buildings, as well as in the courtyards and gardens of wealthier private homes. Typical subjects included the emperors, legendary heroes and animals, as well as outlandish sea creatures, which often formed part of elaborate public fountains. Also very popular were statues that commemorated great victories or mythological scenes.

Many Roman statues were actually copies of earlier classic Greek works (some of which are now known only through the Roman replicas).

An ancient Roman copy of a Greek statue.

Most statues were brightly painted, although the paint has long since worn off surviving examples.

MASS PRODUCTION

Roman sculptors often worked in large workshops that turned out substantial numbers of statues. Many were produced in stock poses to which heads could then be added as desired. Roman statues were more realistic in style than Greek figures. Particularly admired were the busts (head and shoulders) that were produced of the emperors and other notable contemporaries, especially those dating from the early imperial age. These busts probably had their origins in the terracotta busts of ancestors that were made to be displayed at the funerals of Roman aristocrats.

Most sculptures were made to satisfy public demand rather than to pander to the artistic ambitions of the individual sculptors. As a result, relatively few Roman sculptors are now known by name.

Mixing materials

Roman sculptors differed from many others in the ancient world through their mixing of contrasting materials in their work. By using porphyry alongside marble, for instance, the cost of a finished sculpture was substantially reduced.

PAINTING

The ancient Romans were skilled artists. Unfortunately, very few traces remain of their work beyond what has survived in the way of decorated pottery or frescoes.

FRESCOES

The Romans learned the art of wall painting from the ancient Greeks, making their frescoes by painting directly onto plastered walls while they were still wet. Many wealthy Romans hired talented Greek slaves to paint the interior walls of their homes with pleasing scenes. There were several different styles of fresco. The First Style comprised imitations of marble panels, whereas the Second Style consisted of imitations of architectural features, sometimes with small painted scenes framed within painted pillars and similar details. The Third Style included painted scenes of various kinds, while the Fourth Style featured *trompe d'oeil* decoration (subjects painted so as to deceive the eye of the viewer into thinking they are real).

Paint

The paints that were used by Roman artists were made from ground rocks, plant extracts or animal dyes.

This well-preserved Roman mosaic depicts the Titan Oceanus.

MOSAICS

Wealthy Romans often decorated the floors (and sometimes the walls) of their homes with mosaics. These consisted of thousands of tiny cubes (*tesserae*) of tile or stone set in wet plaster to form an attractive geometric pattern or scene. An idea from the East, mosaics had become a popular feature of Roman homes and other buildings by the first century BC.

CRAFTS

Roman craftworkers were skilled in working leather, textiles, wood, metal, glass and pottery, their products being exported throughout the Roman empire and beyond. Some worked in big workshops manned by scores of slaves, while others laboured in smaller workshops in their own homes and sold their goods to passers-by or from stalls in the local forum.

POTTERY

Pottery was a big industry, with thousands of potters, including many slaves and freedmen, producing a wide range of goods, from wine jars and vases to oil lamps and ordinary kitchen crockery. Huge numbers of pots were produced in the kilns of Rome, and archaeological sites are littered with shards of broken pottery. The finest Roman pots were decorated with beautiful scenes painted by skilled artists.

GLASSWORKING

The Romans did not invent glassworking, but did make significant advances in the production of glass objects. Particularly influential was their discovery of ways to make glass objects in large quantities through the use of moulds. Mass production of glass brought the price down, so that glass was no longer a luxury item for the very wealthy. The products of Roman glassworkers included perfume flasks, bottles, jars, beakers and bowls, some of very fine quality. Famous surviving examples of Roman glassware include the Portland Vase (now in the British Museum), which consists of exquisite figures in white glass against a blue background. It is thought to have taken many months to make.

METALWORK

The ancient Romans were skilled metalworkers, making a wide variety of fine objects, from weapons to jewellery, using gold, silver, lead, copper, iron and other metals mined around the empire. They did not know how to achieve a temperature high enough in their furnaces to melt iron, so instead hammered it into shape while it was hot (a process known as forging). They did, however, know how to mix metals into alloys such as bronze (made from copper and tin).

BONEWORKING

Some of the humblest objects found at archaeological sites throughout the Roman empire were made of bone. This cheap but durable material was used for everyday items, such as knife handles, hairpins, combs, needles, gaming counters and dice as well as for sword hilts and as inlay on wooden boxes and furniture.

LANGUAGE

Although the main language of the eastern Roman empire remained Greek (also the preferred language of many Roman scholars), Latin was the everyday language of the Roman world. Latin began as the language of the inhabitants of the plain of Latium, which became the site of the city of Rome. The inhabitants of Rome spoke Latin as they went about their daily affairs, although many other languages were also heard as Roman territory expanded and visitors arrived from distant provinces. Pronunciation varied from region to region,

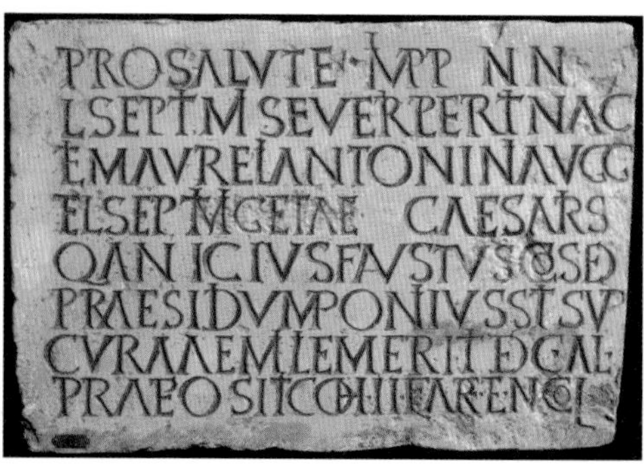

This Roman stone tablet bears an inscription, which is written in Latin.

The Roman alphabet

The Latin alphabet is still used, largely unchanged, for modern European languages The chief difference was that the Romans had only twenty-two letters: i and j were treated as the same, as were u and v, and there were no letters w or y.

but for several centuries Latin continued to function as the main means of communication between well-educated people throughout the known world.

MASTERS OF LATIN

Latin ranked alongside Greek as the language of writers and public speakers and was also used by Rome's administrators. Foremost among acknowledged masters of the language in their own time were the great orator Cicero and the poets Virgil and Martial. The more formal literary Latin of scholars contrasted with the colloquial form (Vulgar Latin) spoken in the streets. It was Vulgar Latin, however, that was to provide the basis of modern Romance languages such as French, Italian and Romanian.

Everywhere the Romans went, they took their language with them, and conquered peoples realized that to deal with Roman administration they would have to learn Latin. The use of Latin throughout the empire led to the disappearance or suppression of many other languages.

LITERATURE

The Romans had a strong literary tradition, building on the example of the ancient Greeks. Many important works have survived through being copied by later scholars. Most Romans were illiterate, and books (in the form of scrolls) had to be copied by hand. Despite this, there were many libraries and bookshops in academic centres such as Rome and Alexandria. The great Roman poets and playwrights were hugely admired and have retained an honoured place in European literature.

CATULLUS

Valerius Catullus was born in Verona around 84 BC and is admired for his witty and elegant verses written in the style of Greek poetry. Of his 116 surviving poems, twenty-five are love lyrics addressed to a married woman named Lesbia. Other poems included accounts of his travels in Asia and criticisms of Julius Caesar and other public figures of the day. He died young around 54 BC.

HORACE

Quintus Horatius Flaccus (Horace), the son of a freed slave, was born in 65 BC. Before establishing a reputation as a poet during the reign of Augustus, he worked as a clerk and fought for Brutus in the civil wars. Success as a poet brought him the friendship of Virgil and Augustus. It raised him from poverty and enabled him to settle in a rural location, which inspired many poems. He left evocative records of contemporary Roman society in such collections of verse as *Odes*, *Satires* and *Epistles*.

JUVENAL

The Roman satirist Decimius Junius Juvenalis (Juvenal) was born around 60 AD. He is thought to have been exiled to Egypt under Domitian because of his critical attitude, but later enjoyed an improvement in status under Hadrian. He is remembered for his *Satires*, written around 98–128 AD. In these, he attacked the corruption and immorality of contemporary Roman society and satirized human weakness in general. He died in 130 AD.

LUCAN

Marcus Annaeus Lucanus (Lucan) was born in Spain in 39 AD. He is remembered for the ten-volume *Pharsalia*, which recounted the civil war between Caesar and Pompey. He committed suicide after being implicated in a plot to kill the emperor Nero.

MARTIAL

The Roman poet Marcus Valerius Martialis (Martial) was born in Spain circa 40 AD. He arrived in Rome around

64 AD and won the support of Seneca and Lucan. He was the author of twelve books of epigrams, containing around 1500 short poems depicting contemporary society in a satirically witty and sometimes lyrical style.

OVID

Born in 43 BC, Publius Ovidius Naso (Ovid) trained as a lawyer before establishing a reputation as a poet. His poetry, mostly based on mythological topics and often on the theme of love, had a profound influence upon later writers. In 8 AD, however, he was banished by Augustus to Tomi (by the Black Sea) and never returned to Rome, dying in exile in 17 AD. His most famous work is the fifteen-volume *Metamorphoses*.

PETRONIUS ARBITER

Petronius Arbiter, a Roman satirist who lived in the first century AD, was famed for his novel *Satyricon*, which comprised a description of the ordinary lives of Romans of different social classes. He committed suicide after being falsely accused of plotting against Nero.

PLAUTUS

Born around 254 BC, Titus Maccius Plautus was one of the most popular Roman playwrights. Just twenty-one of his 130 plays survive, all based on Greek originals, particularly the works of Menander and Philemon. Audiences responded enthusiastically to his often coarse humour and lively use of dialogue, which was later to influence Shakespeare and Molière.

TERENCE

Publius Terentius Afer (Terence) was born in Carthage around 195 BC. He came to Rome as the slave of a senator, who had him educated and granted him his freedom. Of his six plays, only *The Eunuch* enjoyed significant popular acclaim in comparison to the more robust comedies of Plautus. He died in 159 BC.

VIRGIL

Publius Vergilius Maro (Virgil) was born near Mantua in 70 BC. Troubled by the unrest caused by civil war and by the confiscation of his own land, he wrote poems about an idealized pastoral world in the *Eclogues*. His fascination with agriculture and the rural life surfaced again in the *Georgics*. Virgil is most celebrated as the author of the epic twelve-volume *Aeneid*, which took ten years to write and related the adventures of the Trojan hero Aeneas, reputedly the legendary founder of Rome. The Roman emperors, who claimed descent from Aeneas, proclaimed Virgil the foremost Roman writer and his most famous work has remained an object of veneration ever since.

THEATRE

The Romans learned the art of theatre from the Greeks during the third century BC. They built their first theatres in imitation of equivalent Greek structures, initially from wood. Rome's first stone theatre, holding up to 27,000 spectators, was constructed in 55 BC. Numerous structures on a similar scale subsequently appeared throughout the empire.

Roman theatres consisted of a semicircular auditorium, with rows of tiered seating overlooking the acting area. Originally spectators stood, but later rows of stone seats were installed. Spectators brought cushions to sit on, for greater comfort. Different classes of people sat in different areas of the auditorium, the least well-off sitting higher up, while the wealthiest sat nearest the actors. Senators had the best seats at the front. Theatres were open to the sky, but some shelter from the sun was provided by canvas awnings suspended on poles around the auditorium. The design of theatres made it easy for large crowds to enter and leave quickly.

THEATRICAL PERFORMANCES

Actors (who had to be able to sing and dance and play musical instruments as well as deliver lines) performed on a raised platform (the pulpitum) in front of a massive stage wall closing off the open side of the semicircular auditorium. The performers wore masks representing the character they played. These made it easier for the audience to identify who was who at a distance. Most actors (and spectators) were men.

Scenery

The scenery usually consisted of no more than painted cloths hung at the back of the acting area. Scene changes were hidden from the audience by a stage curtain that was raised from a slot at the front of the stage. As the years passed, productions grew more spectacular in order to compete with other forms of entertainment, featuring increasingly imaginative special effects.

Funding

Shows were paid for by wealthy citizens as a means of winning popular support. Tickets were free, but were not easy to obtain due to the popularity of drama, particularly the cruder comedies, with all classes. More accessible for the poor were the simple mimes that actors performed on rough wooden stages in the streets.

Audiences

Roman audiences were very noisy in comparison to audiences today. The spectators clapped, booed, hissed and even rioted while plays were being performed.

MUSIC AND DANCE

Many Roman citizens enjoyed listening to music and watching dancing, although they considered it vulgar and undignified to perform in public themselves, and left this to freedmen and slaves. Music was commonly played in the theatres and at private parties as well as during religious ceremonies and at public events, such as processions and gladiator fights. Followers of some religions, such as the cult of Isis, relied heavily upon the use of frenzied music to work themselves up into a state of religious ecstasy. Professional dancers were often hired to entertain guests at private parties or performed before the crowds in the streets.

INSTRUMENTS

Instruments played by musicians were copied from those of the ancient Greeks. They included lyres, wind instruments (pan-pipes, double flutes and bronze horns) and percussion instruments, such as cymbals,

Military music

Music also played an important role in the armed forces. Orders were often signalled to the legions by blasts on an elaborate spiral horn known as a cornu.

An ancient Roman fresco painting depicting a musical performance.

tambourines and the sistrum (a form of metal rattle). More elaborate was the water organ in which water was pumped into a closed chamber, compressing the air inside, so that when the musician pressed valves air was released into a set of pipes and musical sounds were produced. We know what Roman instruments looked like but have no real idea of the music that Romans listened to as none of it was written down.

PHILOSOPHY

Roman philosophy was mostly an extension of ancient Greek ideas. Particularly influential were the theories that were associated with Diogenes and the class of Greek philosophers known as the Stoics, whose chief principle was an uncomplaining acceptance of situations, good and bad. They also stressed a rational basis (*logos*) for the universe and everything in it. Diogenes himself visited Rome with a group of philosophers and won many admirers for his sober and thoughtful speeches before the Senate.

ROMAN STOICS

Among the most famous Romans influenced by Stoicism was Cicero (106–43 BC), a politician, lawyer and writer. He was widely admired as the greatest public speaker of his day, and played a key role in the events that followed the murder of Julius Caesar, before being murdered by

Victim of the Christians

Hypatia of Alexander (370–415 AD) was a noted and widely respected Neo-Platonic philosopher. After offending the Christian bishop Cyril, however, she was cruelly beaten, tortured and cut to pieces by a Christian mob.

soldiers who were loyal to Mark Antony and Octavian because of his opposition to autocratic rule.

Other Roman Stoics included the Spanish-born writer, philosopher and lawyer Seneca (*c.* 5 BC–65 AD), who served as tutor to the emperors Marcus Aurelius and Nero until he was forced to retire and later to commit suicide. Poets who expressed the ideals of Stoicism in their verse included Lucan and Virgil.

NEOPLATONISM

A more mystical version of Stoicism was Neo-Platonism, which was championed by the Egyptian-born Plotinus (205–270 AD) and other Roman philosophers. Developing the theories of the Greek philosopher Plato, they argued that only by rejecting material nature and cultivating the intellect could the mind enter a perfect state. Among those influenced by Plotinus was the emperor Julian 'the Apostate' who, as a result, sought (unsuccessfully) to return the empire from Christianity to paganism.

Last of the Roman philosophers

Boethius (*c.* 475–524 AD) is sometimes identified as the last of the Roman philosophers. The translator of works by Aristotle, he helped to transmit the philosophical ideas of the ancient world to later generations of thinkers.

HISTORIANS

Although the world's first historians were Greeks, many of the most celebrated Classical historians were Roman.

CAESAR, JULIUS

Julius Caesar (*c.* 100–44 BC) ranks high among Roman historians by virtue of the books he wrote about his military campaigns. Despite a strong propagandist element in his writing, Caesar left a detailed history of his conquest of Gaul in his *Commentarii*.

LIVY

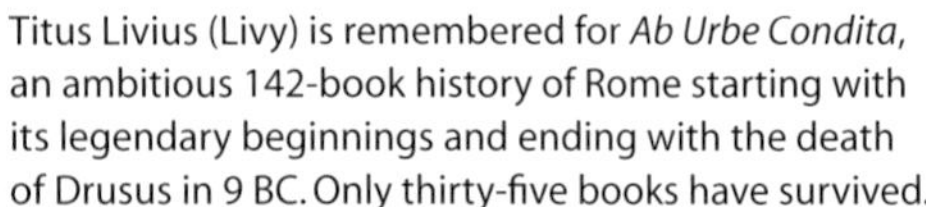

Titus Livius (Livy) is remembered for *Ab Urbe Condita*, an ambitious 142-book history of Rome starting with its legendary beginnings and ending with the death of Drusus in 9 BC. Only thirty-five books have survived.

PLINY

The Roman writer and lawyer Gaius Plinius Caecilius Secundus (Pliny the Younger) was born around 61 AD. He served as consul under Trajan in 100 AD, but is usually remembered for his nine volumes of elegantly written private letters, which offer a unique insight into contemporary affairs. He died around 113 AD.

PLUTARCH

Born in Greece around 46 AD, Plutarch's most important contribution as a historian was his *Parallel Lives*, which comprised biographies of twenty-three pairs of Greek and Roman soldiers and statesmen.

SUETONIUS

Gaius Suetonius Tranquillius was born around 69 AD. As secretary to Hadrian, he was well placed to comment on the great men of his time. He recorded his impressions in such works as *Lives of the Twelve Caesars* and *Lives of Famous Men*, the only examples of his writings to have survived to modern times. He died around 140 AD.

TACITUS

Cornelius Tacitus was born around 55 AD and rose through the ranks of Roman administration to become a consul (97 AD) and governor of Asia (112–13 AD). His influential writings included the historical essays *Germania* and *Agricola* (relating the life of his father-in-law Agricola) as well as the *Histories* and the *Annals*, which comprised full accounts of Roman history in the first century AD. Admired for his concise but evocative writing style, Tacitus believed that the good historian should not take sides, but seems personally to have had republican sympathies. He died around 120 AD.

SCIENCE AND MATHEMATICS

The Romans made many advances in mathematics and science, building upon the discoveries of the ancient Greeks. Many of these had practical applications, as in the construction of large buildings, roads and ships. Some were theoretical speculations on subjects such as the nature of the universe and meaning of existence.

Pliny the Elder (23–79 AD), author of *Natural History*, was notable for his observations of the natural world, while others furthered understanding of geology, mechanics, astronomy, geography, agriculture and medicine. Many Roman thinkers were fascinated by subjects such as astrology and alchemy.

MEASUREMENTS

Some of the most enduring innovations related to measurements. The Roman foot of twelve inches (296mm) has been a standard measurement for centuries, as have the yard and the mile. Also influential was the method of measuring the seasons. The Romans divided the year into the twelve months still used today, each month having three key days: the *calends* (first day of the month), the *nones* (fifth day, or seventh in March, May, July and October) and the *ides* (thirteenth day, or fifteenth in the 'long months' already listed).

ROMAN NUMERALS

The system of numerals was complex, with symbols added together to make larger numbers as necessary. Thus, 2167 in Arabic numerals was MMCLXVII in the Roman system. Roman numerals are still used in certain contexts today, for instance on clock- and watch-faces.

Roman numeral	Modern equivalent
I	1
II	2
III	3
IV (or IIII)	4
V	5
VI	6
VII	7
VIII	8
IX	9
X	10
L	50
C	100
D	500
M	1000

Calculations

The clumsy nature of Roman numerals made mathematical calculations difficult compared to Arabic numerals. Arabic numerals had largely replaced their Roman equivalents by the ninth century AD.

IIII or IV?

The Romans originally wrote the number four as IIII, and they continued to do so even after the introduction of other numerals arrived at through subtraction.

MEDICINE

The Romans relied upon a mixture of science, religion and superstition to treat patients. They based their scientific knowledge of medicine on the writings of the Greek physician Hippocrates and employed many Greeks as doctors. Students at medical schools in Rome were taught a range of skills and received tuition in the subject of anatomy. Others trained in army hospitals, learning from more experienced doctors and surgeons.

Rich people employed personal physicians or paid doctors to visit them at home. Under the terms of the health service established by the Roman state in the first century BC, poorer citizens received treatment free of charge, their doctors being let off payment of taxes in return. In Rome, old or sick people (often slaves) were given shelter at the Temple of Aesculapius on an island in the Tiber. Built after a plague in the third century BC, this became one of the first public hospitals and was a major centre of healing into medieval times.

TREATMENTS

Many complaints, like appendicitis, had no effective treatment and were always fatal. For the treatment of other conditions, however, doctors could choose from various medicines and ointments made from plants, minerals and animal substances. Fenugreek, for instance, was used to treat pneumonia and fennel was believed to calm patients, while eye infections were treated with ointments of lead, zinc or iron.

When conducting operations, surgeons had a range of gruesome iron and bronze instruments at their disposal. These included saws, knives, scalpels, probes, scoops, needles, hooks, spatulas, forceps, speculums and catheters (used to drain bladders). They also used cupping vessels to catch the blood when bleeding a patient to remove the evil 'humour' that they believed was causing the patient's problem. Roman physicians lacked anaesthetics, however, so their patients drank wine to dull the pain of operations.

PREVENTIVE MEDICINE

Because of the risks that were involved in medical treatment, the Romans did what they could to protect their health. They wore rings that were believed to ward off illness and left offerings in the temples of Aesculapius, the Greek god of healing (if treatment proved successful they might leave an offering there in the shape of the relevant part of the body). They realized that diet affected health, and also understood the benefits of exercise, fresh air and regular visits to the baths. Soldiers ate garlic on a daily basis, believing this would protect their health.

PART NINE

Rome today

Impressive ruins of the ancient Roman civilization may still be seen in the city of Rome itself and elsewhere throughout the territories that were once part of the Roman empire. Archaeologists continue to make remarkable discoveries, while many cities still enjoy the benefits of Roman roads, bridges, aqueducts and sewers. Even more significant, however, has been the legacy of Rome in relation to the languages, legal systems, arts and sciences of the modern world.

Ruins of the Roman Forum
The streets and squares where ancient Romans walked may still be seen in modern-day Rome itself, and in many other places around the former empire.

DISCOVERING ANCIENT ROME

Sources of information about ancient Rome range from the writings of Roman historians, such as Tacitus and Suetonius, and commentators, such as Juvenal and Martial, to the discoveries of modern archaeologists at sites as far removed as northern Africa, the Middle East, Britain, Gaul and Rome itself.

ARCHAEOLOGY

Finds unearthed by archaeologists have ranged from temples, villas and bath-houses to smaller but equally significant objects, such as statues, pottery, weapons, jewellery and coins. In many places, Roman walls and other structures remain in evidence in the lower sections of medieval defences or other buildings. Elsewhere, archaeologists have dug beneath modern buildings to expose the foundations of much earlier Roman towns.

Preservation

Some Roman temples and other buildings owe their survival to their later use as Christian churches. The Curia in Rome's Forum (in which the Senate met) served from the seventh century AD as the church of St Hadrian.

Rescue archaeology

Many recent discoveries have been made in the course of 'rescue archaeology', in which sites are investigated (often hastily) by archaeologists before being built on or developed.

SIGNIFICANT FINDS

Roman ruins may still be seen throughout Western Europe and the Middle East. Some have survived in a more or less intact state because of their remoteness. Yet others have been preserved against all odds in the midst of modern cities, often after being protected under a build-up of debris over the centuries. The Colosseum in Rome, the Arena in Verona and the cities of Pompeii and Herculaneum are good examples.

What you can see today

We can still wonder at the sophistication of Roman building and engineering at protected sites through what was formerly Roman territory. Among the scores of amphitheatres, temples and villas, there remain such marvels as the Colosseum in Rome, the Pont du Gard in southern France – a three-storied aqueduct, 50 kilometres (30 miles) in length, that brought water to the city of Nîmes – and Hadrian's Wall in Britain. In some places, such as Pompeii or Thamugadi (Timgad in Algeria), one can view the remains of whole towns or cities.

ROME

At the heart of Roman history and culture was the Forum of the city of Rome, substantial ruins of which can still be inspected today. Amongst the rubble may be seen significant archaeological remains, such as the Curia (where the Senate met), the triumphal arches of Septimius Severus and Titus, the temples of Castor and Pollux, Antoninus and Faustina, Vesta and Vespasian and the Via Sacra (Sacred Way), the route that triumphal processions took up to the Capitoline Hill. On the Palatine Hill overlooking the Forum may be found the ruins of imperial palaces.

Nearby stand the Colosseum, Trajan's Forum (dating from the second century AD and including Trajan's Column, carved with scenes of the emperor's military triumphs), the remains of Nero's Golden House, the Arch of Constantine and other Forums, as well as the Mamertine Prison, which once housed the captive Gaulish chieftain Vercingetorix and St Paul.

Other monuments

Elsewhere in Rome are monuments such as the Pantheon (a perfectly preserved classical temple with what was, for centuries, the widest masonry dome in Europe), public bath complexes and the remains of the vast stadium known as the Circus Maximus.

POMPEII AND HERCULANEUM

South of Rome may be seen the unique sites of Pompeii and Herculaneum. These two Roman towns were buried under ash when the volcano of Mount Vesuvius exploded without warning one summer afternoon in August 79 AD. The population was caught completely by surprise, and many people were buried with their homes and all the details of everyday life in up to four metres (thirteen feet) of ash and pumice.

EXCAVATION

The sites were first excavated in the nineteenth century, when archaeologists found much material in a near-perfect state of preservation. The towns have proved rich sources of information about the lives of ordinary

Bodies

The most moving of the artefacts found in Pompeii and Herculaneum are the plaster casts of the bodies of those who died when the volcano exploded. Although the bodies rotted away long ago, they left an impression in the ash that settled around them: these hollows were later filled with plaster by archaeologists, thus recording the dead at the moment they were entombed.

Romans, as revealed from the remains of shops, homes and streets. Archaeologists have also found some inscriptions, mosaics, wall paintings and graffiti, as well as personal artefacts such as shoes, and perishable food items such as bread, eggs and figs. In one house, they even found water in a kettle waiting to be boiled.

These plaster casts record the exact positions of Romans killed when Vesuvius erupted in 79 AD.

ROMAN BRITAIN

Britain, which marked the northernmost edge of the Roman empire, has a number of notable archaeological sites dating back to the time of the Roman occupation. New villas and other buildings are located each year, and many modern roads still follow routes laid down by Roman engineers. The most remarkable sites include Bath, Fishbourne, Hadrian's Wall and St Albans.

Bath

The Celts were the first people to establish the natural hot spring at Bath as a shrine. They dedicated it to their goddess Sulis. When the Romans arrived in Britain, they developed Bath (or Aquae Sulis) as a spa town dedicated to Minerva, constructing large temples and bathing complexes. These fell into disrepair after the Romans left, but they were substantially restored and enlarged in the eighteenth century. The baths are well-preserved, although everything above the level of the pillar bases is of post-Roman origin.

Fishbourne

A large Roman villa, often described as a palace, was discovered at the village of Fishbourne, near Chichester in West Sussex, in 1960. It has been dated to the time of the Roman invasion of Britain in 43 AD and was abandoned after a fire towards the end of the third century AD. The palace is notable for its extremely fine mosaic floors.

Hadrian's Wall

Hadrian's Wall ranks among the most well-known of all Roman remains. Stretching 130 kilometres (80 miles) from the Solway Firth on the west coast to Wallsend on the east coast, it was built on the orders of the emperor Hadrian by three legions between 122 AD and 129 AD, to act as a barrier against the warlike tribes of Caledonia and also to provide a base from which the Roman army could control everything to the north.

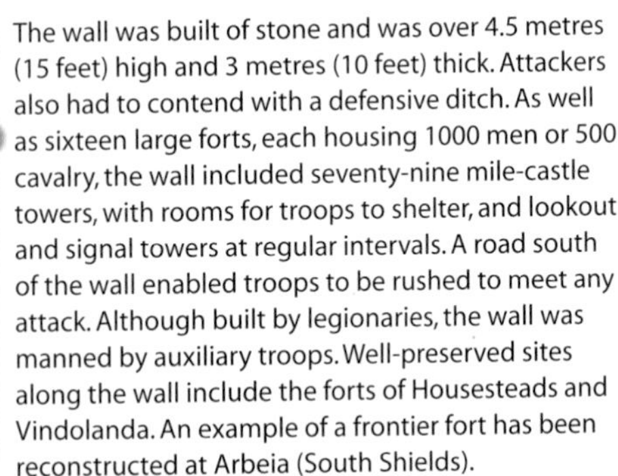

The wall was built of stone and was over 4.5 metres (15 feet) high and 3 metres (10 feet) thick. Attackers also had to contend with a defensive ditch. As well as sixteen large forts, each housing 1000 men or 500 cavalry, the wall included seventy-nine mile-castle towers, with rooms for troops to shelter, and lookout and signal towers at regular intervals. A road south of the wall enabled troops to be rushed to meet any attack. Although built by legionaries, the wall was manned by auxiliary troops. Well-preserved sites along the wall include the forts of Housesteads and Vindolanda. An example of a frontier fort has been reconstructed at Arbeia (South Shields).

Hadrian's Wall was the most substantial of all the walls built by the Romans to defend the empire's frontiers.

St Albans

St Albans (or Verulamium) was one of the largest towns in Roman Britain. Substantial ruins remain of not only an amphitheatre but also of defensive walls and the extensive foundations of baths, temples, shops and residences. Numerous finds from the area are housed in a museum in the town.

THE CULTURAL LEGACY

Interest in ancient Roman civilization revived during the Renaissance of the late fifteenth century. Roman objects were collected and put in museums where scholars and artists might study them. The Roman style of architecture was copied throughout England and France, giving birth to the so-called Classical style. The study of Latin texts (the Classics) became a central part of European scholarship, and many aristocrats embarked on a 'grand tour' of archaeological sites. The publication of Edward Gibbon's *The Decline and Fall of the Roman Empire* (1776–88) promoted further interest.

Roman influence

The influence of the ancient Romans may be detected today in a wide range of contexts and in many nations of western Europe and the Mediterranean basin. Their ideas (extending those of the ancient Greeks) provided the foundation of many fields of scholarship, from law and government to town planning and the arts. The continuing influence of ancient Rome is evident even in the way we speak; thousands of words in modern European vocabularies are of ultimately Latin origin. Many Latin words and phrases are still in common use, often in untranslated form, among speakers of English, as revealed by the selection opposite.

Latin word or phrase	Meaning
ad hoc	improvised
ad infinitum	indefinitely
ad lib	spontaneously
affidavit	statement under oath
agenda	list of things to be done
alias	assumed name
amen	so be it
bona fide	in good faith
campus	university site
carpe diem	seize the day
caveat emptor	buyer, beware
circa	around
compos mentis	of sound mind
contra	against
credo	belief
cum laude	with honours
curriculum	programme of study
de facto	actual
ego	self
ergo	therefore
et cetera	and so on
exempli gratia	for example (e.g.)
exit	way out
facsimile	copy
factotum	servant
focus	centre of attention
gratis	free
habitat	natural surroundings
homo sapiens	human being
ignoramus	idiot
index	list of references
in extremis	in difficulty
in loco parentis	in the parents' place

Latin word or phrase	Meaning
in memoriam	in memory of
in vitro	(with)in the glass
magnum opus	masterpiece
mausoleum	large tomb
maximum	greatest amount
mea culpa	I am to blame
media	means of communication
memento	souvenir
minimum	least amount
minor	lesser
modus operandi	method of operation
nausea	sickness
per se	in itself
plaudit	praise
posthumous	after death
prima facie	on the face of it
pro	for
pro rata	in proportion
quasi	resembling
quid pro quo	something for something
quorum	necessary number of people
quota	proportional part
re	concerning
rostrum	platform
rota	list or roster
sanatorium	medical centre
tacit	unspoken
terra firma	firm ground
ultimatum	final demand
verbatim	word for word
versus	against
vice versa	the other way round
vox populi	general opinion

MUSEUMS

Fascinating and often beautiful artefacts from ancient Rome are preserved today in museums around the world. Many collections were first assembled in the eighteenth century and continue to grow as more relics are uncovered by contemporary archaeologists.

Among the most important collections in the city of Rome itself are those of the Museo Nazionale Romano, which are housed at five separate locations, such as the Baths of Diocletian. These include many fine examples of Roman art, including mosaics and sculptures (among them the famous second-century AD marble copy of Myron's Discus Thrower). Also worthy of note are the collections at the Vatican, Musei Capitolini, Palatine Museum and Antiquarium and at the Villa Giulia.

Other museums

Elsewhere in the world, major collections of Roman art and antiquities are held in both local and national museums. The collection of Roman artefacts at the British Museum in London is particularly noted for its collections of Roman glass and silver. There are also interesting collections of Roman objects at many other locations in the UK, including museums at Caerleon, Cambridge, Chester, Corbridge, Colchester, Maryport, Newcastle, Oxford, Silchester and York.

FIND OUT MORE

BOOKS

Baker, Simon, *Ancient Rome: The Rise and Fall of an Empire*, 2006

Boardman, J., Griffin, J., and Murray, O., *The Roman World*, 1988

Chevallier, R., *Roman Roads*, 1976

Clare, John D., *I Was There: Roman Empire*, 1992

Cornell, T., and Matthews, J., *Atlas of the Roman World*, 1982

Deary, Terry, *The Rotten Romans*, 1994

Gibbon, Edward, and Trevor-Roper, Hugh, *The Decline and Fall of the Roman Empire*, 1994

Goldsworthy, Adrian, *Caesar: The Life of a Colossus*, 2006

Goldsworthy, Adrian, *The Complete Roman Army*, 2003

Goodman, Martin, *The Roman World*, 2004

Heather, P.J., *The Fall of the Roman Empire: A New History*, 2006

Holland, Tom, *Rubicon: The Triumph and Tragedy of the Roman Republic*, 2004

Hornblower, Simon, and Spawforth, Antony, *The Oxford Companion to Ancient Civilization*, 2004

James, Simon, *Eyewitness Guides: Ancient Rome*, 1990

Jones, Terry, and Ereira, Alan, *Terry Jones' Barbarians*, 2006

Marks, Anthony, and Tingay, Graham, *Romans*, 2003

Matyszak, Philip, *Chronicle of the Roman Republic*, 2003

Rodgers, Nigel, *Roman Empire*, 2006

Scarre, Chris, *Chronicle of the Roman Emperors*, 1995
Suetonius and Graves, Robert, *The Twelve Caesars*, 2003
Virgil and West, David, *The Aeneid*, 2003

WEBSITES

The following is a brief selection of websites about ancient Rome:

www.bbc.co.uk/schools/romans
(general information for children)

www.crystalinks.com/rome.html
(general information)

www.fr-novaroma.com/Archeology
(informal guide to recent ancient Roman finds)

www.historyforkids.org/learn/romans
(general information for children)

www.historylink102.com/rome
(general information)

www.thebritishmuseum.ac.uk/world/rome/rome.html
(guide to the Roman collections in the British Museum)

GLOSSARY

Amphitheatre Semi-circular stadium for entertainments.
Amphora A two-handled jar with a narrow neck.
Aqueduct A structure built to convey water in pipes over valleys and rivers, etc.
Asia Minor That part of the Middle East closest to European borders.
Atrium The central area of a Roman house.
Augury Ritual divination of future events by a priest.
Augustus A title borne by the Roman emperors.
Aureus A Roman gold coin.
Auxiliary A soldier who is not a Roman citizen.
Ballista A Roman catapult.
Barbarian A person from beyond Roman borders.
Bestiarius A handler of wild beasts in the Roman arena.
Bulla A charm worn to ward off evil spirits.
Caesar A title borne by the nominated heir to the imperial throne.
Caldarium A hot room in a bath complex.
Capital The head of a pillar or column.
Carthaginian Of or relating to the civilization of Carthage, in north Africa.
Catacomb An underground cemetery.
Celtic Of or relating to Iron Age peoples of northern Europe.
Censor A civil magistrate of ancient Rome.
Centurion A military officer in charge of a century.
Century A military unit of 100 (later, 80) soldiers.
Citizen A person with full legal and voting rights.
Cohort A tactical unit of Roman soldiers.
Consul One of two senior government officials, elected annually by the Senate.
Corinthian An order of classical architecture, characterized by columns with capitals decorated with carved acanthus leaves.

Crucifixion A method of execution in which victims are left on a cross.
Decimation The military punishment of executing every tenth man in a legion.
Denarius A Roman coin.
Domus A Roman house.
Doric An order of classical architecture, characterized by columns with undecorated capitals.
Elysium The Roman heaven.
Equestrian A senior officer in the Roman republic.
Etruscan Of or relating to the Etruscan tribe of central and northern Italy.
Fasces A symbol of authority comprising an axe bound in a bundle of rods.
Forum An open space in a town or city, surrounded by administrative buildings.
Freedman A former slave awarded his freedom.
Frigidarium A cold room in a bath complex.
Gaul The Roman name for what is now France and part of Germany.
Gens A clan formed by related families.
Gladiator A fighter in the Roman arena.
Goth A member of the Germanic peoples who invaded Italy in the late fifth century AD.
Grammaticus A teacher.
Hun A member of the nomadic tribes that overran much Roman territory in the fifth century AD.
Hypocaust An underfloor heating system.
Insulae An apartment, or block of apartments.
Ionic An order of classical architecture, characterized by columns with scroll capitals.
Lararium A shrine dedicated to household gods.
Latin The language of the ancient Romans.
Legion A military unit of as many as 5000 soldiers.
Legionary A soldier in one of Rome's legions.
Lictor An official who bore the fasces before important Roman magistrates.
Ludus An elementary school.
Lyre A musical instrument.
Maniple A military unit of 120 soldiers.

Numen A guardian spirit.
Paedagogus A slave who accompanied the children of wealthy Romans at school.
Patrician A member of the Roman aristocracy.
Pax Romana The lengthy period of peace that existed under Roman imperial rule.
Peristyle A row of columns around a temple.
Pilum A heavy javelin carried by Roman soldiers.
Plebeian A member of Rome's urban poor.
Pontifex Maximus The High Priest of Rome.
Praetor A senior official of the Roman republic.
Praetorian Guard The imperial bodyguard.
Proconsul A former consul who was entrusted with the governorship of a province.
Province A territory outside Rome under Roman control.
Quaestor A senior financial official.
Quincunx Military formation.
Quinqereme A warship with five banks of oars.
Sabines A tribal people of central Italy.
Senate The ruling council of Rome.
Sestertius A Roman coin.
Taberna A Roman tavern.
Tartarus The Roman hell.
Tepidarium A warm room in a bath complex.
Terracotta A brownish-red pottery in building and crafts.
Tesserae The tiny cubes used to make a mosaic floor.
Tetrarchy The sharing of the rule of the empire by four men.
Toga A robe worn by Roman men.
Tribune An official appointed to represent plebeians in the Assembly.
Trireme A warship with three banks of oars.
Triumvirate The sharing of the rule of the empire by three men.
Vandal A member of the Germanic tribes that sacked Rome in 455 AD.
Vestal Virgin A priestess who served in the temple of the goddess Vesta.
Viaduct A road-bearing bridge.
Villa A large house in the country.

INDEX

Look out for further titles in the Collins Gem series.